Photo courtesy of Jack Meagher

Alasdair McGregor was born in Sydney in 1954 and graduated with honours in architecture at the University of New South Wales. Since the early 1980s he has devoted himself almost exclusively to painting and later to writing and photography.

In 1983 he co-organised the Heard Island Expedition and has since visited the sub-Antarctic islands and the Antarctic continent several times. He has also travelled extensively in the Pacific and Australia, especially in the Northern Territory and the Kimberley, continuing a commitment to his interpretation of the region's natural landscapes.

Alasdair McGregor has participated in twenty group exhibitions and has held thirteen solo exhibitions. He has completed many corporate and public commissions both in Australia and overseas and his paintings are represented in many public and private collections.

He is the author, with Quentin Chester, of *The Kimberley: Horizons of Stone* (1992) and *Australia's Wild Islands* (1997).

For Charmian and in memory of Ian.
May there always be 'books, music, love and laughter'.

We dwelt on the edge of an unspanned continent, where the chill breath of a vast, polar wilderness, quickening to the rushing might of eternal blizzards, surged to the northern seas. We had discoverd an accursed country. We had found the Home of the Blizzard.

Sir Douglas Mawson
The Home of the Blizzard

mawson's huts

An Antarctic Expedition Journal

Alasdair McGregor

For Sheila,
A 'little Antarctic Story'
very best wishes,
Alasdair

HALE & IREMONGER

Sydney
March '05

Photograph on title page: Ninnis and a dog team haul supplies up to Aladdin's Cave. (Douglas Mawson, Mitchell Library, State Library of NSW)

10 9 8 7 6 5 4 3 2 1

Typeset by
Midland Typesetters Pty Ltd
Maryborough, Victoria 3465

Printed and bound by
Southwood Press Pty Ltd
80–92 Chapel Street, Marrickville, NSW 2204

For the publisher
Hale & Iremonger Pty Ltd
PO Box 205, Alexandria, NSW 2015

National Library of Australia
Cataloguing-in-Publication entry:

McGregor, Alasdair, 1954– .
Mawson's huts: an Antarctic expedition journal

Bibliography:
ISBN 0 86806 653 2.

1. McGregor, Alasdair, 1954– Diaries. 2. Huts–Antarctica–Denison, Cape–Conservation and restoration. 3. Historic sites–Antarctica–Denison, Cape–Conservation and restoration. 4. Antarctica–Discovery and exploration. 5. Cape Denison Historic Site (Antarctica) –Description and travel. I. Title.

919.89

This project has been assisted by the Commonwealth Government through the Australia Council, its arts funding and advisory body.

Colour section designed by Cassi Design

Foreword

Heroes are not too common in this day and age, and outstanding heroes like Sir Douglas Mawson demand our admiration and respect.

His Main Hut at Cape Denison is a memorial to a great Australian and should, I believe, be restored to its original condition.

I have been delighted to read in this diary of the success of the first AAP Mawson's Huts Foundation expedition, which has ensured the immediate survival of the hut and begun the task of conservation of the whole of the historic site at Cape Denison, where Mawson and the men of his Australasian Antarctic Expedition (AAE) landed in 1912.

I recall vividly my visits last year to the splendidly restored Antarctic exploration huts of Captain Robert F. Scott and Sir Ernest Shackleton. We flew by helicopter to Captain Scott's hut at Cape Evans, which is in excellent condition after considerable maintenance by the New Zealand Antarctic Restoration Trust. It was impossible not to be impressed with this journey back into history—back into the heroic age of Antarctic exploration.

We then flew north to Cape Royds, to Shackleton's hut. I had always admired Ernest Shackleton more perhaps than any other Antarctic explorer. He was renowned for his courage and skill in moments—or months—of unimaginable stress and danger. I had never been in his hut before and to go inside was an eerie experience. I had the distinct sensation that I was meeting face to face my most revered Antarctic hero. I had never had such a sensation before and it was one I will long remember.

The re-establishment of the Cape Evans and Cape Royds huts has, I believe, been a very important historical event in the Antarctic. As more and more tourist vessels sail the Antarctic waters, hundreds of people are being given the experience of visiting, in carefully controlled conditions, these relics of a bygone age.

Last year I sailed south in the 20,000 tonne *Marco Polo* with a restricted passenger list of 400 people on board—mostly of middle age or indeed elderly. We cruised off Cape Evans while our ten inflatable Zodiac craft took load after load ashore. The passengers had been warned that if they had any physical problems it was recommended that they stay on board ship. But not one of them did—the lame, the ancient and the decrepit all went ashore to gaze in awe at the contents of Scott's hut and then returned to the ship feeling they had just had a unique experience.

Mawson's AAE was the first Australasian expedition to a completely uncharted area of Antarctica and his achievements rank him without question as one of Australia's greatest explorers. That is why it is so important that those frozen pieces of history which immortalise Mawson and his team be restored.

Alasdair McGregor, author of this diary, is a modern-day explorer who captures with his sketch-book and the camera lens, as well as in his journal, the excitement, the grandeur, the awesome scenery and the dangers and privations that are so much a part of Antarctic travel.

This is a remarkable record of an important project, which like Mawson's own expeditions, was privately organised and funded, and which adds immensely to our knowledge of the Cape Denison environment and defines the spirit of the place.

Sir Edmund Hillary, KG, KBE
July 1998

Introduction

Inching towards an anchorage near the coast of Antarctica on 12 January 1997, the Russian polar-research vessel *Akademik Shokalskiy* was treated to conditions that would have been hailed as memorable by any Bondi beach-goer or Thredbo ski-buff. But the postcard-perfect weather gracing that day belied the true nature of the place. We had arrived at Commonwealth Bay on the coast of George V Land—'the home of the blizzard'—perhaps the most wind-harried and inhospitable place on earth.

It was at Cape Denison—a low and largely ice-free peninsula probing the broad indent of Commonwealth Bay—that 85 years before, geologist and academic, Dr Douglas (later Sir Douglas) Mawson stepped ashore from the converted sealing vessel SY *Aurora* on 8 January 1912. He was accompanied by a party of young men comprising the Australasian Antarctic Expedition (AAE).

Nearly three months after Mawson's landing and more than fifteen hundred kilometres to the south-east, Captain Robert Falcon Scott and the last of his companions died on their desperate retreat from the South Pole. That familiar saga will be forever part of Britain's imperial folklore. Mawson had been invited on that ill-fated journey, and had he not been of an independent mind, brimming with self-confidence, he would probably have died with Scott. Then his only claim to posterity might well have been as a footnote to Scott's tragic folly and the glorification of heroic failure.

Born in Yorkshire in 1882, Mawson migrated to Australia with his parents at the age of two. Settling in Sydney, the young Mawson showed early academic promise at Fort Street High School and at the age of sixteen entered the University of Sydney where he studied mining engineering and geology. By 1905 he had degrees in both disciplines and in the same year joined the staff of Adelaide University. Even before Antarctica figured in his interests, Mawson had undertaken fieldwork in remote localities. While still at the University of Sydney he participated in a geological survey of Vanuatu and in 1906 Mawson discovered the first significant deposit of uranium in outback South Australia.

When Mawson landed at Cape Denison he was already hardened to the rigours and privations of polar exploration, having served with Ernest Shackleton's 1907–09 British Antarctic (*Nimrod*) Expedition to the Ross Sea. Not only had he been among the first party to climb the near 4,000-metre active volcano, Mount Erebus, he had also participated in the longest unsupported sledge journey yet made in Antarctica. Together with fellow Australian geologist and mentor, Professor T.W. Edgeworth David, and Dr Alistair Mackay, a Scottish surgeon, Mawson journeyed from Cape Royds, McMurdo Sound to the South Magnetic Pole. (Mawson had been included in the *Nimrod* Expedition at David's behest.) The party, in the preferred British tradition of the time, man-hauled their sledge more than two thousand kilometres to the Pole and back, claiming the region for king and empire at the farthest reached point.

When Mawson returned to Australia in 1909, the frozen south was still veiled in mystery. Little more than a fraction of its coastline had ever been sighted, let alone explored. The first human footprint had only been left on the continent in 1895 and Amundsen's conquest of the South Geographic Pole still lay two years off.

Almost directly south of New Zealand at 170°E lies Cape Adare, at the head of the vast bight of the Ross Sea. The region to the cape's south was one of the first parts of Antarctica to attract explorers, starting with James Clark Ross in the 1840s. As interest in reaching the South Geographic Pole began to mount, the Ross Sea, and the gigantic expanse of the Ross Ice Shelf at its head, were seen as convenient shortcuts to the interior of the continent. By the early years of the new century, the traffic to the Ross Sea, if not exactly busy was concentrated in more or less the same direction. To the west of Cape Adare, in the sector directly to the south of Australia, almost nothing was known of the Antarctic coast. The German *Gauss* Expedition had spent the winter of 1902 beset by pack ice near the Shackleton Ice Shelf, drifting from about 96°E to 80°E, but from there eastwards, the only knowledge stemmed from two intrepid voyages made more than sixty years before. The American, Charles Wilkes, explored this stretch of coastline in 1840, and although some of his claims of sighting land have been disputed, he determined the general limits of the coast for more than a thousand kilometres. In the same year, the French explorer, Jules-Sébastian-César Dumont d'Urville, discovered land deep in the Southern Ocean in the course of a sweeping ethnographic and scientific investigation of the Pacific. Dumont d'Urville landed on an offshore islet about a hundred kilometres west of Commonwealth Bay in January 1840 and named the surrounds Terre Adélie for his wife.

Back in Australia after the *Nimrod* Expedition, Mawson's passion for exploration did not remain sated for long. In the preface to *The Home of the Blizzard*, his popular account of the AAE, Mawson states:

> . . . once more a man in the world of men, lulled in the easy repose of routine, and performing the ordinary duties of a workaday world, old emotions awakened, the grand sweet days returned in irresistible glamour, far away voices called . . .

In the following year, 1910, Mawson visited England and Europe. There, the idea of exploring the stretch of Antarctica that he referred to as his 'Land of Hope and Glory' really started to take hold. But it's obvious that the notion had been seeded in his mind sometime earlier, as he admits in *The Home of the Blizzard*:

> At times, during the long hours of steady tramping across the trackless snow-fields, one's thoughts flow in a clear and limpid stream, the mind is unruffled and composed and the passion of a great venture springing suddenly before the imagination is sobered by the calmness of pure reason. It may have been in one such phase that I suddenly found myself eager for a glimpse of the great span of Antarctic coast lying nearest to Australia.

While in England, Mawson petitioned Scott to extend the scope of his forthcoming *Terra Nova* Expedition to encompass exploration west of Cape Adare. He was unwilling, but attracted by the young veteran's undoubted talents and physical strength, Scott offered Mawson a berth on the *Terra Nova*. Moreover, he proposed that Mawson be in his polar sledging party. Fortuitously, the offer was declined. Mawson's interests outshone pure geographical conquest. He also revealed his scheme to Shackleton. So enthused by the idea was Mawson's former leader, that he proposed to take charge of the expedition himself with Mawson as his subordinate. This offer was also declined by Mawson, who then resolved to organise the expedition on his own terms. And so the AAE was born.

> For many reasons, besides the fact that it was the country of my home and Alma Mater, I was desirous that the Expedition should be maintained by Australia. It seemed to me that here was an opportunity to prove that the young men of a young country could rise to those traditions which have made the history of British Polar exploration one of triumphant endeavour as well as of tragic sacrifice.

Back in Australia, Mawson wasted no time in soliciting support from the influential and the wealthy. He obtained the backing of the Australasian Association for the Advancement of Science and lobbied governments, both directly and through his growing band of supporters. In early 1911, Mawson again visited England. There, he finalised the purchase of the 600-tonne *Aurora*—which, at the time he still didn't have the funds to pay for—garnered donations from wealthy Australians living abroad and persuaded the British Government to assist the AAE. By the time the expedition departed Hobart on 2 December 1911, Mawson had already proven himself to be a consummate organiser and leader.

To be able to explore as much of Antarctica directly to Australia's south, Mawson planned to establish three separate coastal bases on the continent and a fourth, primarily as a radio relay station, on sub-Antarctic Macquarie Island. After successfully landing a party of five with enough supplies for a year's sojourn on Macquarie, the *Aurora* pushed further south toward an alien coast. The grossly overloaded ship floundered slowly onwards through heavy seas, finally reaching the edge of the pack ice on 29 December. The expedition had already travelled past the edge of the then known world.

Mawson was blessed to have as his ship's master and second-in-command, 28-year-old Captain John King Davis. Like Mawson, Davis had served with Shackleton and brought to the expedition the experience and cool-headed judgement needed to negotiate uncharted waters, strewn with ice. For the next ten days, Davis gingerly threaded the *Aurora* around and through the pack, as the ship slowly probed westward. No land was sighted until 6 January. Mawson was running out of time. The expedition had already been forced well west of its proposed initial landfall of about 158°E, and still no suitable landing site had been found. Finally, on 8 January 1912, the *Aurora* steamed into a broad ice-free bay, where:

Mid-winter's dinner, 21 June 1912. (Photo: Frank Hurley. Courtesy of Mitchell Library, State Library of NSW.)

> The land was so overwhelmed with ice that, even at sea-level, the rock was almost entirely hidden. Here was an ice-age in all earnestness; a picture of northern Europe during the Great Ice Age some fifty thousand years ago or more.

Roughly centred in the bay, a small rocky peninsula was sighted and late in the afternoon, a whale boat was dispatched to investigate:

> We were soon inside a beautiful, miniature harbour, completely land-locked. The sun shone gloriously in a blue sky as we stepped ashore on a charming ice-quay—the first to set foot on the Antarctic continent between Cape Adare and Gaussberg, a distance of about two thousand miles.

Bare rock on which to found a hut lay just 50 metres from the shore and the adjacent glacial ice offered an easy prospect for sledging parties running to and from the polar plateau rising directly behind. This had to be it. Over the next ten days, supplies and equipment sufficient for several years sojourn were landed on the shores of the peninsula that Mawson named Cape Denison, after a Sydney benefactor of the expedition.

Workshop under construction, January 1912. (Photo: Frank Hurley. Courtesy of Mitchell Library, State Library of NSW.)

With the summer now running desperately short, Mawson abandoned plans for three field parties and combined the two eastern groups at Cape Denison. On 19 January, Davis and the *Aurora* departed, intent on establishing a westerly station under the leadership of Frank Wild, yet another veteran of Shackleton's *Nimrod* Expedition. Four weeks from Cape Denison, and after a desperate search for a suitable landing site, Wild's party was dropped on the Shackleton Ice Shelf off Queen Mary Land, nearly twenty-five hundred kilometres to the west of Commonwealth Bay. It was a calculated risk, establishing a base on the floating ice shelf. However, the location proved to be quite sound and by the time of their relief in February 1913, Wild and his seven men had explored nearly five hundred kilometres of difficult coastline.

While the *Aurora* crept westward, battling gales and heavy pack ice, Mawson and his party set about immediately establishing their winter quarters. The larger of two living huts was made ready for occupation by the end of January. A second hut, originally intended as shelter for the second eastern party was then abutted to the leeside of the first. The larger hut was to be used as general living quarters and the smaller as a workshop, laboratory and radio room.

The huts had been designed by Mawson. Both were prefabricated in Australia and shipped to Antarctica in broken-down form, ready for quick site assembly. They were framed in Oregon timber. The larger was clad in tongue-and-groove Baltic pine boarding, while the smaller was sheeted in boards of an as yet unidentified species of pine, possibly radiata. The larger hut, with an internal floor area of 53.5 square metres was manufactured by George Hudson in Sydney and donated to the AAE. The smaller, measured just under twenty-seven square metres and was purchased from an Adelaide merchant. There were no internal divisions in either hut, apart from a small partitioned cubicle that served as Mawson's bedroom and a photographic darkroom adjacent the stove and kitchen in the living section.

In addition to its internal living and work spaces, the combined hut was surrounded on all but the leeside of the workshop by enclosed verandahs up to 1.8 metres wide. The verandahs were intended for storage of food and other supplies and for housing the party's nineteen Greenland husky sledge dogs. Each of the roofs was continuous over its verandah and extended to within a couple of metres from the ground. Additional boxes of stores were stacked around the weather sides of the hut, so as few vertical surfaces as possible were presented to the wind.

By 30 January, eighteen men had taken shelter in a building about half the size of a small suburban house. Outside, atrocious winds blew with mounting ferocity as the summer waned, and all too quickly the party realised that the local geography conspired against them. They had inadvertently built their home in a raging river of wind. The already furious winds draining uninterrupted off the polar plateau accelerated at a terrifying pace as they funnelled down the broad valley behind the camp. With the onset of winter, the winds grew worse and the cold more intense. During May 1912, the average wind speed for Cape Denison was recorded at just over forty-seven knots. Other months were almost as bad. Recent studies have confirmed the AAE's wind speed data and concluded that Cape Denison is not only the windiest place in Antarctica, but also the windiest place close to sea level on the entire planet.

Mawson was just 30 when he landed at Cape Denison and most of his party were still in their twenties. Many were fresh from studies at Australian and New Zealand universities. Their roles were to be many and varied, including those of geologist, magnetician, biologist, cartographer, meteorologist, dog handler, taxidermist, storeman and astronomer. Among the group was Frank Hurley, a 24-year-old photographer from Sydney. The AAE was to be the first great adventure for Hurley in a remarkable career that saw his own fame come to almost rival that of Mawson's.

On 17 January 1912, while the AAE were busy establishing their winter quarters at Cape Denison, Robert Falcon Scott and his four companions, turned for the long march back from the South Pole. Defeated and broken, robbed by Amundsen of what Scott referred to as 'the reward of priority', within two and a half months they were all dead. Unlike Scott, Mawson's objectives and motivation had always been more intellectual than visceral. The AAE came not to conquer Antarctica but to explore and to learn. As a creature of his time, Mawson did have imperial and geographical imperatives in his sights, but these were secondary to the quest for knowledge. Science was seen as a sufficient goal in itself, in the areas of geology, biology and physics.

However, Mawson was also mindful of the resources of the unknown land and its surrounding seas. He foresaw the possibility of economic minerals or guano deposits. Also, the sealing and whaling potential of the Southern Ocean was still largely untapped, and if the region were to be exploited meteorological data needed to be gathered.

Three smaller huts (the Transit and Absolute Magnetic Huts and the Magnetograph House) were built in the early months of 1912 to house experiments or serve as scientific observatories, but most importantly the main hut stood firm, providing a base from which much was achieved in the name of science by Mawson and his young party. Through the bitter winter, the hut was both home, haven and workplace. The men rode out the worst weather Antarctica could possibly hurl at them, spending their days in the enforced comradeship of the hut, while preparing for spring and summer sledging trips. The hut became the ultimate sanctuary, as Mawson noted in his diary after one particularly violent storm:

> There is no exaggeration in stating that probably every other hut that has previously been erected in the Antarctic would have been carried

> away bodily long ere this ... Imagine the awful end of any other expedition having come to this spot with the huts provided them ... Imagine a hut with all in it suddenly blown bodily into the sea.

Mawson and his men made their first tentative exploratory forays in September and October, before sledging parties were finally dispatched in early summer along the coast and hinterland to the west and east, and inland towards the South Magnetic Pole. Each party faced the same pitiless winds and cold. Danger shadowed them always, and hunger and exhaustion were often their unwanted companions. But for Mawson's own far eastern party, the journey was marked by calamity.

On 14 December, 500 kilometres from winter quarters, the unthinkable happened. Belgrave Ninnis, one of Mawson's two companions, fell to his death along with a fully laden sled and dog team, when a snow bridge covering an enormous, gapping crevasse suddenly collapsed. Among the cargo on Ninnis's sled was most of the party's food. Mawson and his remaining companion, Xavier Mertz, then began a desperate forced march back to Cape Denison. To supplement their meagre rations, the pair killed and ate the remaining dogs as each became too weak to pull their sledge. Mertz managed to get within 160 kilometres of their objective before dying of exhaustion, starvation and vitamin A poisoning, brought on by eating dogs' liver. Mawson struggled on alone, debilitated by the same toxic effects that had contributed to Mertz's death, and finally made it back to Cape Denison on 8 February 1913, eight weeks after Ninnis had died. His journey across the savage wilderness of the Antarctic ice cap is perhaps the greatest feat of lone survival in polar history. Mawson stated in his diaries upon reaching safety:

> What a feeling of gratitude to Providence for such deliverance. I had intended to push on to the utmost in the hope of reaching a point where my remains would be likely to be found by a relief expedition, but I had always hoped against hope for more. Now I had arrived at the goal of my utmost hopes. The outlook was so different that I was overcome with a smooth and soft feeling of thanksgiving.

Mawson had just missed relief by the *Aurora*. Totally debilitated by his ordeal, he was then forced to endure a second winter at Cape Denison. He was not alone though, as five of his colleagues had remained behind in the hope of the far eastern party's return. A scientific program proceeded during a winter even more savage than the first, and a sledge journey was made in late November. The *Aurora* returned to Commonwealth Bay for a third time in December 1913, and after a month-long exploratory cruise of the Antarctic coast, she turned north for Adelaide and a hero's welcome. Of the welcome home, Mawson wrote, 'the voice of the innumerable strangers—the hand-grips of many friends—it chokes one—it cannot be uttered'. Mawson went on to a distinguished career in teaching and research. In 1921, he became Professor of Geology and Mineralogy at the University of Adelaide, a post he held for the remainder of his career.

The achievements of the AAE rank among those of the greatest of polar expeditions. An aggregate of over sixty-five hundred kilometres of the Antarctic coast and hinterland was negotiated by sledging parties from both continental stations, providing a wealth of geographical information. Extensive biological and geological collections were made on the continent and on Macquarie Island. The *Aurora* engaged in detailed oceanographic research, both inshore and in the deep ocean. Magnetic records were continuously amassed at the closest station to the South Magnetic Pole yet maintained, and wireless communication to and from Antarctica was pioneered. The AAE's achievements provided the initial authority for later territorial claims by Australia to more than 40 per cent of the Antarctic continent. Mawson further explored the Antarctic region, leading the British Australian and New Zealand Antarctic Research Expeditions (BANZARE) of 1929–31. In two lengthy summer cruises, huge tracts of the continent's coastline were claimed by Mawson for the British Empire and by implication, Australia.

Landing at Cape Denison on that uncharacteristically still and warm January evening in 1997, my first impressions of Mawson's winter quarters were disarming. Nestled close by the shore, the main hut's placement seemed overwhelmingly appropriate. With graceful, twin pyramidal roofs and a short side fronting the prevailing wind, it was hard to imagine that grand natural landscape without such a defiant reminder of human endeavour. From a distance the hut's pine cladding gleamed in the dipping sun. It looked so fresh, perhaps almost new, not a structure that had withstood all the assaults of nature for more than eighty years. Yet from closer quarters, clues as to the true nature of that hostile place were plainly etched into the fabric of the hut.

Paradoxically, the extremes of the Cape Denison climate have helped to both preserve and inexorably destroy much of the fabric of the building. Abandoned after the AAE, the main hut began to fill with wind-driven snow and ice which consolidated as a frozen mass, anchoring the structure to its foundations. But the same wind-driven ice particles have seriously abraded the external cladding, with many boards now worn perilously thin. The building's fresh appearance is simply graphic evidence of a process identical to sand blasting.

I made that brief, January 1997 visit to Cape Denison in the company of Geoff Ashley, conservation architect and timber-building specialist with the Sydney firm of heritage consultants, Godden Mackay. After several hours detailed survey, Geoff's preconceptions were confirmed. He concluded that it was perhaps a matter of only a few years before the weakened cladding boards were broken by the force of the wind and torn from the building. The hut would then descend into ruin and eventual obliteration. The workshop seemed particularly vulnerable. Its boards were originally thinner than those on the living hut and Geoff found them to be very close to failure. They originally measured sixteen millimetres in thickness, but Geoff's checks on the workshop roof found that in places up to a third of the cladding's depth had been stripped away by the wind and ice. Many boards were now only ten to twelve millimetres thick. Of the three scientific observatories, one was in remarkably good condition, but the other two were now in ruin. Also, a memorial cross erected by the AAE to Mertz and Ninnis was found to be in need of repair.

Despite the best of intentions and no lack of enthusiasm on the part of the few government or private parties to visit Cape Denison over the two decades prior to our visit, very little had been achieved towards the preservation of Mawson's Huts.

An Australian National Antarctic Research Expeditions (ANARE) party in 1978 was the first to effect any repairs to the main hut. It excavated accumulated snow and ice from inside the building, in particular the workshop. The party also made external repairs using lead sheet, canvas tape and silicone. All of these remedies have proved totally ineffectual in preventing the building from again filling with ice and snow. In the mid 1980s, Project Blizzard, a dedicated group of enthusiasts and adventurers, mounted two expeditions to Cape Denison. As the buildings are listed on the Register of the National Estate, Project Blizzard was somewhat restricted in the scope of the repairs it was permitted to carry out by the Australian Heritage Commission. However, it did thoroughly survey all the Cape Denison structures and inserted temporary props under collapsing ceiling members in the living section of the main hut.

All grand schemes rise in the minds of individuals with foresight. As Mawson alone was the wellspring of inspiration behind the AAE, so the vision of a coordinated effort to conserve and care for the Cape Denison buildings rests with one man. David Jensen is corporate affairs director for the media and communications company, Australian Associated Press (AAP). Originally from New Zealand, David had long been impressed by the New Zealand Government's restoration and stewardship of the other four surviving huts from the 'heroic' era of Antarctic exploration. Scott's Huts at Cape Evans and Hut Point and Shackleton's at Cape Royds, plus Carsten Borchgrevink's 1899 hut at Cape Adare all lie within New Zealand's Antarctic claim, the Ross Dependency. Since the late 1950s, New Zealand has been progressively restoring and caretaking the huts, conserving a wealth of artefacts and monitoring the effects of tourist visitation in what has become a busy corner of Antarctica. Since 1987 the work has been coordinated and promoted by the Antarctic Heritage Trust.

David Jensen increasingly felt that if it was good enough for New Zealand to look after the huts within its domain then it was beholden upon Australia to do likewise with its own Antarctic relics. With the backing of his company he established the AAP Mawson's Huts Foundation in 1996 with the express purpose of raising funds for the conservation of Australia's Antarctic heritage. It was an appropriate project for AAP, as they had been providing daily news bulletins to Australia's Antarctic stations for 30 years. Like Mawson, Jensen began lobbying government for financial support and sought sponsorship in kind from Australian companies, some of whom had been backers of the AAE. He coopted the help of the Australian Antarctic Division, which has international responsibility for historic sites within Australia's territorial claims, and obtained the patronage of the Governor–General, Sir William Deane.

I met David Jensen through a mutual friend in late 1996. At that time Antarctica was the farthest thing from my mind. It was ten years since I'd been south, so I assumed that Antarctica was a cherished but complete chapter of my life. I'd been co-organiser and participant in the 1983 Heard Island Expedition and had made brief visits to the Antarctic continent and Macquarie Island in the mid 1980s with ANARE. Heard Island, in particular, was a revelation and a true turning point in my life.

When I encountered David, my work was already well known to AAP. I'd completed a major painting commission for the foyer of their Sydney building in 1994, and the company had purchased several works from an exhibition of mine on the theme of Australian islands. David saw the potential of having an artist and photographer with some Antarctic experience involved with the fledgling foundation. Architectural training was also a bonus. He generously offered me a berth on a tourist ship bound for Antarctica in December of 1996—almost on the spot! In return, I pledged some of the expected paintings to the Foundation to assist with its fundraising and agreed to document the trip photographically. The ship was the *Akademik Shokalskiy*, a Russian vessel under charter to a New Zealand tourist operator, Southern Heritage Expeditions.

In the course of its voyage, the *Shokalskiy* was to call at a number of sub-Antarctic islands south of New Zealand, then proceed to Commonwealth Bay and, weather permitting, land its passengers at Cape Denison for a brief but privileged glimpse of Mawson's Antarctica. Southern Heritage Expeditions had offered two berths to the Foundation, so that it could visit the huts and make an up-to-date assessment of their condition. Geoff Ashley's firm was engaged to provide that assessment and so the two of us boarded the *Akademik Shokalskiy* in the New Zealand port of Bluff at the southern end of the South Island on 27 December 1996. Over the next week or so we pitched and rolled south, enjoyed good food and company, and called at spectacular and wildly beautiful islands.

But as the *Shokalskiy* pressed further south we shared a growing sense of unease. What if the notorious Cape Denison winds prevented us from getting ashore? The ship wouldn't wait indefinitely for an inspection of the huts to be made. There were other destinations on the tourist itinerary that had to be satisfied. We could've been going to Antarctica for no good purpose, our only consolation being an enforced holiday. As events transpired, there were no problems getting ashore and we saw Cape Denison just as Mawson had first encountered it. However, our shared anxiety on the southward voyage served to highlight in one small way the difficulties in successfully caring for Mawson's Huts.

Cape Denison's extreme remoteness, far from Australia's usual

Antarctic operations, plus the equally formidable difficulties of effecting practical, sensitive and durable repairs in such a hostile environment all militate against the preservation of the huts. The closest Australian Antarctic station to Cape Denison is Casey, about thirteen hundred kilometres to the west. The mere thought of dispatching a ship just to Commonwealth Bay and the formidable impact on our already stretched Antarctic budget that would entail makes government planners cringe. That's where David Jensen and the Mawson's Huts Foundation come in. By collaborating with private tourist operators like Southern Heritage, Jensen saw the potential to get shipping costs—a large part of any Antarctic operation—within manageable limits.

Of course, the visit that Geoff and I made to Cape Denison in early 1997 was just the briefest of preludes to the main event. Basically, all we had done was confirm the belief that repairs to Mawson's Huts were urgently needed. Jensen and the Foundation had already chartered the *Shokalskiy*, in collaboration with Southern Heritage, for two voyages to Commonwealth Bay and return, the first in December 1997 and the second several weeks later. Berths were to be allocated to a conservation work party and the remainder sold to tourists who would travel on each of the round trips.

Throughout 1997, the Foundation continued to lobby and cajole. The Antarctic Division became increasingly involved, loaning the Foundation everything from clothing and cookware to tents and sleeping bags. Rob Easther, an experienced member of the Division's staff (by coincidence a friend of mine from Heard Island days), was appointed to lead the expedition in the field and applications for volunteer team members were invited through the media. Midway through the year Rob had to withdraw from the expedition due to an impending addition to his family, but he continued in the vital role of overall project manager. His place in the work party was taken by Alan Grant, a Brisbane-based landscape architect, surveyor and experienced Antarctic hand. Alan had spent 1993 as leader at Mawson Station. Geoff Ashley devoted much of 1997 to formulating a conservation plan in consultation with the Australian Heritage Commission and the Antarctic Division, and was appointed conservation team leader and deputy field leader for the forthcoming expedition.

As well as Geoff, Alan and myself, three builders, an archaeologist, a cook and camp manager, doctor, communications operator and an AAP journalist were all appointed to the expedition. At the eleventh hour, investor finance was raised by a Western Australian film production company and a two-man documentary crew joined our group, taking our complement to thirteen in total.

In mid December 1997, I found myself heading to Antarctica for the second time in a year. My role was to act as official photographer, with a brief to document everything from daily life to work in progress to archaeological finds and the Antarctic landscape. I was also going to paint, and so continue to help the Foundation through exhibitions and the like upon my return to Australia.

I left behind a household in a state of mild turmoil. Major renovations to my home and studio were well advanced but still several months from completion. And through all the chaos of building and preparing for Antarctica, I was yet to finish a group of five large paintings commissioned by a client in Asia earlier in 1997. The building work would continue in my absence in the capable care of my very tolerant and supportive family; and the last of the paintings would just have to wait. Like Mawson, the thought of the 'grand sweet days returned in irresistible glamour' was too much for me. I was off to Antarctica, regardless.

Not just symbols of national achievement, Mawson's Cape Denison Huts are recognised in the Antarctic Treaty—the international agreement endorsed by all nations with interests in the south—as priceless and rare remains from the heroic age of Antarctic endeavour. The value of preserving them as monuments to intelligent enquiry, resourcefulness, tenacity, innovation and sheer courage is, I believe, beyond question.

It might be argued that the main hut in ruin would be no less a monument. Before my first visit to Cape Denison I gave that idea some credence. However, the intact form of that tenacious little building surviving against all odds is surely a far more powerful embodiment of the spirit of Mawson and the AAE than any ruin. In ruin, that spirit and its symbolism would be sadly diminished. As global travel and communication becomes increasingly easy, so remoteness is extinguished. At least at Commonwealth Bay that remoteness is still guarded emphatically by the elements, making Mawson's Huts of ever increasing and more universal value.

The huts stand at the end of the great age of European exploration and its concomitant four centuries of colonial expansion. But not only do they mark the limits of one era, they presage another:

that of our own contemporary world, the age of technology. Seventy-five years before Mawson's birth, no human being had ever seen the seventh continent. Mawson was already twelve years old when the first human footprint was left in the snows of the Antarctic mainland. But by the time of his death in 1958, Antarctica had been fully explored and mapped; and for the last time on a continental scale, the unknowable had become the known. This rapid expansion of geographic and scientific knowledge was in large measure the result of the vision of a handful of men such as Douglas Mawson. As symbols of that vision, Mawson's Huts are far more than mere historic relics. They are an embodiment of a pivotal time in the growth of human understanding of our planet. This is the story of their rescue from oblivion.

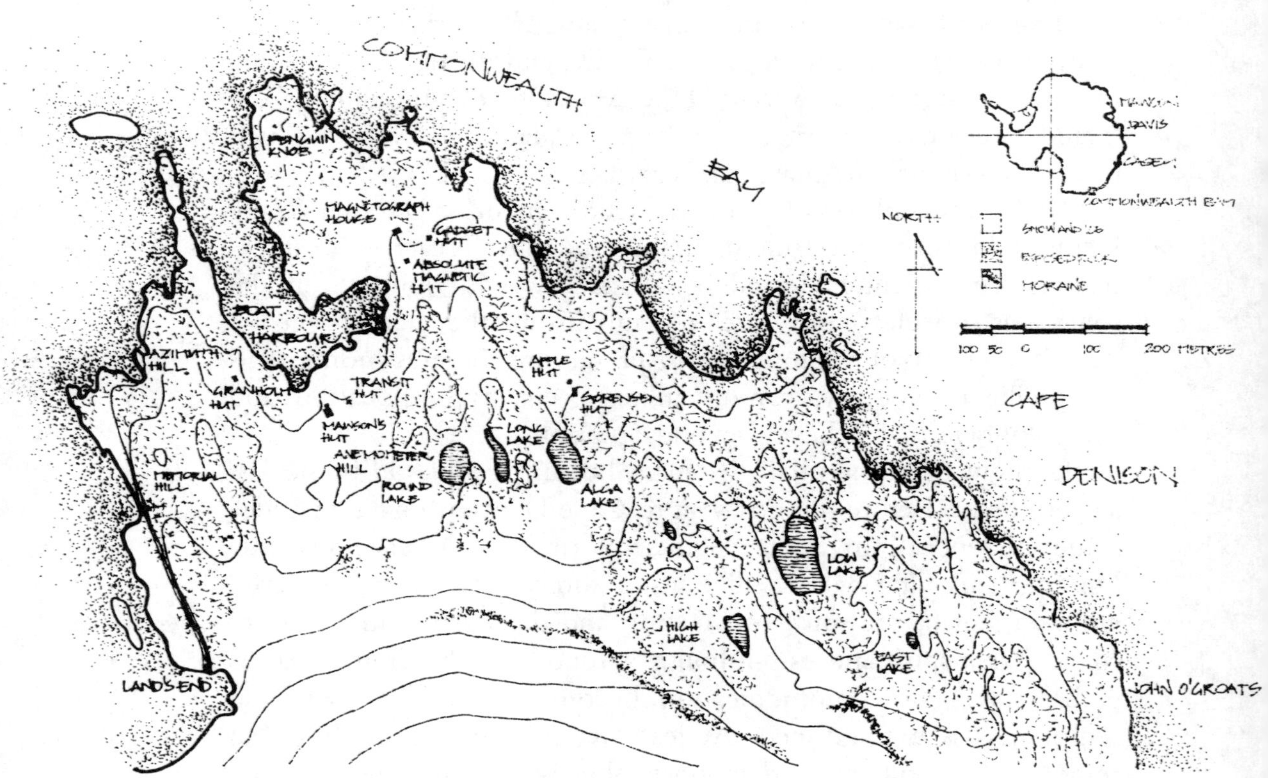

Early Days

The *Akademik Shokalskiy* is set to leave Hobart for Antarctica on the morning of Thursday 11 December 1997. For three days prior to departure our group will spend most of its time in last-minute packing and preparation at the Australian Antarctic Division's headquarters at Kingston, just south of Hobart.

It's good to be back in Tasmania. Since I was 21, this place has never been far from my life. I feel strongly that had I not spent time here in my formative years, I may well not be heading for Antarctica now. It all began on a holiday visit to Tasmania in 1975. As a student of architecture, part way through my studies, I was enthralled by the island state's historic charms: its Georgian cottages, country gardens and fine public buildings. But nothing had prepared me for the tug of Tasmania's wilder parts. Here I saw great forests, dripping and dark, and of its remoter places, I kindled the thought of sights never before seen by anyone. One day stands out among many memories of that visit, still bright after more than twenty years.

While not an instant revelation, that day confirmed a growing sense within me. Driving towards Queenstown, I think, the road was flanked by a ragged saw-blade of grey granite rising from its forest mantle. A boisterous stream raced under the bridge we had just crossed, then disappeared among meadows of ochre-coloured button grass and copses of sage-green woodland. I had to stop. Down off the verge, the button grass disguised the inevitable Tasmanian mud but I didn't have to go far to realise that I had to make such magnificent wild places an integral part of my life. And so, in the 20 or so years since, I've been lucky enough to do more or less just that, with the last ten years being more rather than less.

But what has all this to do with Antarctica? My interests did shift, confirmed by the opportunity to spend the following year working in Tasmania. Painting, photography and ultimately writing about wild places followed. After my year in Hobart, I returned to architectural studies in Sydney; but not long after finishing university, I found myself participating in an expedition that voyaged deep into the Southern Ocean, to Heard Island. I've never recovered from that experience: with glacier-clad Heard Island, I caught the Antarctic 'disease'.

There's something appropriate too about Tasmania being both the bridge by which I crossed to an alternative way of life, and the embarkation point for my next visit to Antarctica. I'm made aware of this parallel in one of those strangely out-of-context, yet pleasant interludes in the whole pre-departure scramble. On the day prior to our sailing, the group is diverted from the organised chaos and overall-clad activity of the Antarctic Division store to the refinement of neo-Gothic Government House and lunch with the State Governor, Sir Guy Green and Lady Green. Hobart glides through a balmy summer's day. Lavender-fringed fountains gurgling above a sweep of lawn and formal gardens flowing down to the Derwent River seem remote from any connection with Antarctica.

However, over lunch, Sir Guy gives a speech which eloquently details the historic links between Hobart and Antarctic exploration. He explains that Dumont d'Urville, Scott, Amundsen and Mawson had all used Hobart as a staging post to Antarctica or first port of welcome return. The farewell parties, Sir Guy notes, 'were legendary'. It was from Hobart in 1912, that Amundsen sent secret word to the King of Norway of his polar triumph, not allowing his civilisation-starved men off their ship until confirmation of his message had been received from home. Government House itself had a connection, Sir Guy tells us. James Clark Ross set up a magnetic observatory, 'Rossbank' in the vice-regal grounds in 1840, one of a network of stations, to assist in his search for the South Magnetic Pole. With the talk of these great expeditions, the fine china and glassware, and panelled and gilded ceilings embellished with coats-of-arms, all seem so distant from the Antarctic Division store, let alone the weeks to come.

At Kingston, our sorted and packed cargo steadily grows. Two semi-trailers will be needed to transport it all to the ship. Building materials, a gas stove, tents, fuel bottles, shovels, mattresses, rescue equipment, pots and pans, water containers, a wheelbarrow, radios and a four-wheel all-terrain motorbike, known as the 'quad' are all gathered, ready for transport. There's even a vacuum cleaner! Why is this commonplace domestic appliance destined for Antarctica? The driest and windiest continent on earth is probably also the least dusty. But amidst the frantic activity of the last few days, I'm content to let such questions drift. Someone surely has more than fastidious housekeeping in mind.

One of the largest cargo items will be the timber. New members have been purchased for intended structural repairs to the huts, and a quantity of Baltic pine boarding is sorted, given a coat of timber preservative, then bundled and wrapped. Much of it will go as deck cargo on the *Shokalskiy*. The Baltic pine is well seasoned, having sat in the Antarctic Division's store for nearly twenty years. It was purchased subsequent to the ANARE work party at Commonwealth Bay in 1978, but has never quite made the voyage south. There seems almost enough material to build a new hut, so structural timber and cladding not used by our party will be stockpiled at Cape Denison for future use.

And then of course there's the food—enough dried and packaged provisions to sustain thirteen people for at least ten weeks, plus emergency supplies if the unthinkable happens. While no one seriously doubts that our stay in Antarctica will be but a relatively short one, being marooned for anything up to a year is a disturbing, if unlikely, contingency. Add to this the perishable and frozen supplies to be loaded directly on board the ship, together with sixty cartons of beer and a liberal quantity of wine, and one starts to wonder about the more immediate difficulties in actually landing all this impedimenta in Antarctica.

It's difficult not to think about what it'll be like struggling with full gas bottles for instance, each weighing about a hundred kilograms, as they're slung from the ship's crane into a pitching inflatable boat. Then there's the trip to shore through near-freezing waters, not once but dozens of times. I suppose mild comfort can be drawn from the fact that our load, although daunting, will be nothing like the 150 tonnes of stores that the *Aurora* carried south.

Small and loose items are stowed in modular plywood boxes, each one carefully packed to avoid shifting and breakage of its contents at sea. The packed boxes then receive a brief scrawled description on their lids, and a slap of paint: yellow for food, green for hardware and tools, red for kitchen and blue for technical and scientific. Large and bulky items end up in steel-mesh caged pallets, that'll go directly into the ship's hold, and more delicate equipment plus some of our mountain of personal gear is crammed into a small upright shipping container known as a 'WOV (wanted on voyage) box'. The WOV box will be lashed to the deck of the *Shokalskiy*.

Over three days, the packing and sorting proceeds apace, under the steady guidance of Rob Easther and Alan Grant. I'm struck by the attention to detail required. Going back months, Rob, in consultation with David Jensen, Geoff Ashley and others, has seen that our needs have been identified, quantified and the goods sought, either from sponsors or the Antarctic Division. What couldn't be begged or borrowed has simply been bought.

The Antarctic Division has supplied the expedition with its specialist clothing, most of which was issued on a previous visit to Hobart. However, there are a few adjustments and returns to be made as most are not convinced by the 'one size fits all' theory. As the expedition is receiving sponsorship, our distinctive yellow windproof jackets, known as 'ventiles' are all returned temporarily to the store and there emblazoned with embroidered patches showing our supporters' logos: Digital, Canon, Jaeger, Yalumba, Lanier, Sydney Allen Printers, Avis, and, of course, AAP itself.

Everyone pitches in and we're helped by a keen group of friends and Division staff. Most of us were tired before we arrived but, despite the effort, we all enjoy the chance to work together for the first time and get to know each other a little better before departure.

'Departure' is a word that has ruled the lives of the thirteen expeditioners since the team was finalised about six months ago. All are busy, successful people in their own workaday worlds and are coming together because of a passionate 'want' rather than any need. As all but Alan, Geoff, Dave Killick (our AAP journalist) and the film crew are volunteers, the high measure of commitment among the group is underscored by varying degrees of sacrifice. Everyone has extended themselves to be here. We've come to Hobart from all over the country and are now thrust into each other's company for better or worse for the next two months.

We all know basically what to expect, and while this is a very special project, for some, going to Antarctica is as familiar and welcome as going home. Most are in their forties, considerably older than Mawson's team. While we might lack youthful exuberance, we're certainly not short of Antarctic experience. In all, the party boasts fifteen winter sojourns and many more southern summers between us. Only for three, will Antarctica be a totally new revelation.

About two thousand five hundred Australians have wintered with ANARE since the first stations were established in the late 1940s but the 'Antarctic family' is still a small one. So, while we've all come from different directions to join the expedition, a number of our team are old friends and work mates.

Geoff and I have already weathered one voyage together and think we can tolerate each other in the course of a second. We're both from Sydney, as is Dr Estelle Lazer our archaeologist. Estelle lectures at the University of Sydney and knows what to expect from Commonwealth Bay, having accompanied Project Blizzard's first team in the mid 1980s. She assures me that the weather only gets worse the longer one stays at Cape Denison. We both share a love for yet another windswept and wild place—Heard Island, of course. Estelle has conducted archaeological investigations of the nineteenth-century sealing industry there. To both of us, it's pretty close to the most spectacular spot on the planet.

Alan Grant our leader is from Queensland, as is the first of our two Daves. David Gillott, building contractor, farmer and aquaculturalist from North Tamborine on the Gold Coast hinterland, has one full year plus a second Antarctic summer behind him. Dave has already stamped himself as the most energetic of the group. He seems to have two speeds: asleep and full velocity.

Nearly half of our contingent is from South Australia, which, I suppose is appropriate in the Mawson context. From Adelaide there's Joan Russell, our cook and camp manager. In real life, Joan is the recently appointed personnel director for the South Australia Police, the most senior civilian member of the service. She's also no stranger to Antarctica, having served twice as an ANARE station leader. Improvising with food and cooking under difficult conditions will be nothing new to Joan, as she was once a patrol officer's wife in Papua New Guinea.

Also from Adelaide is Dr Rod Givney, our medical support. Rod's well known to Joan having been medical officer at Casey in 1990, under her leadership. The pair remain good friends. Although Rod has trained and practiced extensively in emergency medicine, his main interests these days lie in infectious diseases. As he concedes, small isolated communities in Antarctica are probably the least fruitful ground imaginable for the practice of this branch of medicine. Quiet and thoughtful, Rod's concern for our individual and collective welfare is displayed in his admission that he'll be very nervous for the duration of the expedition. He knows, and I think we all understand, that in the event of a serious medical emergency, very little effective treatment would be possible. At a team get-together a few months ago at Cradle Mountain in Northern Tasmania, Rod said 'as long as all I have to do is be nervous and shovel snow, I'll come home happy and very relieved'.

The third of our South Australians is builder Paul Delaney from Mylor in the Adelaide Hills. Paul's spent several seasons in Antarctica, most recently as maintenance foreman at Mawson Station over an eighteen-month stretch ending in April 1997. Like Dave Gillott, Paul is also a small farmer and by coincidence the two worked together at Davis Station in the summer of 1993–94. Paul has 'almost' been to Cape Denison in the past. He was part of an ANARE team that was to land there from the supply ship *Lady Franklin* in 1988 to consolidate a recently established summer base near Mawson's Huts. Heavy pack ice prevented the *Lady Franklin* from getting anywhere near Commonwealth Bay and the team returned to Hobart with building materials that will again be shipped south, this time on the *Shokalskiy*.

With the possible exception of Paul, our communications person, Allen Rooke from Wooli on the New South Wales north coast, makes the Antarctic credentials of the rest look flimsy. 'Rookie', as he seems to be known to all, has spent a total of six winters and nine summers in the south and has been awarded the Antarctic Medal for his longstanding service. Rookie's radio experience goes all the way back to his days as a boy sailor in the Royal Australian Navy in the 1960s. Like Rod, Rookie also knows Joan well, having been her deputy station leader at Macquarie Island in 1994. His radio and communications expertise aside, Rookie was one of the last dog handlers in Antarctica and accompanied the final dog team to be removed from the Australian stations on their journey to new homes in North America.

Our building team is completed by Gerard 'Ted' Bugg, a ranger from the Cradle Mountain–Lake St Clair National Park in Tasmania. This will be Ted's first trip south, but what he lacks in Antarctic experience is amply compensated elsewhere. A carpenter and joiner and one-time mineral prospector, Ted's work duties include the care and maintenance of a string of historic timber huts in the Cradle Mountain park.

Journalist, Dave Killick is also an Antarctic novice. He vied with dozens of other applicants within AAP to win the Mawson's Huts assignment, and so is as keen as the rest of us. Like any good journalist Dave is versatile, having covered everything from the Sydney crime beat to the Atlanta Olympics. At 29, Dave is also the youngest member of our party by more than ten years. Working in collaboration with Rookie, it'll be his task to send back news stories, digital pictures and daily diary entries for the Foundation's website—all via satellite. Many features of our Cape Denison experience won't be dissimilar to Mawson's. That can't be said of communications.

And watching, filming and recording every exciting or trying moment will be cameraman Malcolm Ludgate and producer-cum-sound recordist Mike Piper. Together, they take the South Australian contingent to five. For Mike this will be his first trip south but for Mal, Antarctica is familiar ground. The Mawson's Huts expedition will be his seventh summer in the south. Two of Mal's previous trips have been to shoot footage for the award-winning Imax spectacular *Antarctica*. Both Mike and Mal have a special interest in wildlife film making—everything from seals to seahorses to echidnas. While we're in Antarctica, Mike will have a fixed camera set up for the duration in a goanna's burrow, somewhere in the South Australian desert, trying to record the illusive lizard's breeding behaviour.

Just a few days ago, we all came together in Sydney for the first time in several months. On a hot and muggy Saturday afternoon we gathered in the gracious harbourside surrounds of Admiralty House for an official farewell by the Mawson's Huts Foundation patron, the Governor–General, Sir William Deane and Lady Deane. Sir William's heart-felt words of encouragement made us all feel very special. He, like so many others, was particularly interested in how I planned to paint under the extremes of the Antarctic climate. Alas, all I could proffer for the time being was what I suspected sounded like unconvincing theory.

Sweat trickled down my back as the talk turned to the faint veil of smoke dimming the afternoon glare. The bushfire season had already begun. El Niño, the 'child' had arrived early—a Christmas offering unwanted by all. Out on the lawns sweeping down to the harbour, young magpies strutted, proclaiming their growing stature among the family group. Thousands of kilometres to the south, at Cape Denison, young Adélie penguins would soon be struggling against their unforgiving surroundings for an equally precious, single chance at survival.

The last day in Hobart is spent in final packing, checking and a few mad scrambles to buy previously overlooked items, both communal and personal—everything from camera batteries and motor oil to ear plugs. We also continue discussions with the Antarctic Division personnel relevant to our own areas of interest or responsibility. The Division, through the manager of their Multimedia Unit, Kevin Bell, generously offers 50 rolls of colour transparency film in addition to the 100 or so already supplied by our film sponsor, Fuji. Kevin sees our visit as a great chance to remedy the paucity of images of Cape Denison and the huts held in the Division's photographic library.

Almost in a daze, I find myself that evening at a farewell dinner at Hobart's Grand Chancellor Hotel. Speech follows speech. It's all very flattering and genuine, if not a little embarrassing, as we're extolled as heroes in the making by more than one speaker. While the accolades flow, somewhere out in the darkness of the nearby Derwent River, the *Shokalskiy* has berthed. As we enjoy dinner and mingle with our twenty or so fellow passengers for the first time, two semi-trailer loads of cargo are being crammed into the ship's tiny hold.

The Journal

Thursday, 11 December

That day is finally here. The expedition has been accommodated in style at the Grand Chancellor during our last night on stable ground. I've shared with Dave Killick, and like all the rooms facing east, ours commands imposing views of the Derwent River and the Hobart wharves. On a distant pier, I can just make out a ship with a white superstructure and mid-blue hull. Sharing the same berth, and making the first ship look quite small, is a sleek, almost sinister-looking black craft, that I'm pretty sure is the new Bass Strait catamaran, the *Devil Cat*. All Tasmanians seem excited about this new link with the mainland.

The smaller ship, I know to be the *Akademik Shokalskiy*—she's quite familiar to me now. Built in Finland in 1983, the Russian-flagged *Shokalskiy* and its several sister ships, started life as oceanographic and fisheries research vessels, but are now mostly under charter for tourist cruises. Being ice-strengthened, they are particularly suited for high latitudes. From its home port of Vladivostok, the 72-metre *Shokalskiy* sails under a Russian crew of 24 and when cruising accommodates 38 passengers.

But if the *Shokalskiy* looks small near an ocean-going ferry, the AAE's *Aurora* would be overwhelmed. At only 50 metres in length, the ex-New-Foundland sealer, *Aurora*, was barely large enough for Mawson's needs. But with sturdy frames and planking, and an inclined bow that could bear down on and crack the pack ice, the *Aurora* was an ideal vessel for the Antarctic seas. Of equal importance was the addition of sails, allowing coal consumption to be minimised. Mawson saw sails as a '*sin qua non* in Polar navigation . . . a last resort should the coal-supply be exhausted or the propeller damaged'.

A jazz band entertains the growing crowd of well-wishers who have turned out on yet another sparkling Hobart morning. Reporters scribble notes and camera crews (including Mike and Mal) crowd around our spokesmen, Alan, Geoff and Rob, while in the background the last of the luggage is bundled on board. Most of the bags arrived earlier from the hotel. To my surprise and relief, two large backpacks, a kit bag, camera cases, tripod, portfolio, and sundry smaller packages are waiting for me on my bunk. Geoff and I will once again put up with each other and share a cabin. It'll be a bit cosy, as Geoff has a similar quantity of gear, but it won't be anywhere near as squashed as Mike and Mal's cabin. Sharing the space with even just a fraction of their most precious equipment, it looks as though it'll be hard enough to open the door, let alone get into bed. The *Shokalskiy*'s cabins are quite comfortable and spacious, and if not luxurious, they're more than adequate. Above all, they'll be warm, dry and private retreats when the going gets rough. I'm prone to sea sickness as are a number of us. Although not as bad as some, I expect to become very familiar with my bunk over the next few days.

The WOV box is lashed on the upper aft deck and two large bundles of timber have been secured to the ship's foredeck. The rest of our cargo, apart from personal items and perishables, has somehow found its way into the *Shokalskiy*'s small hold.

Compared to the comfort offered on ships of Mawson's era, the *Shokalskiy* is a palace. On its southward voyage, the grossly overloaded, narrow-beamed *Aurora* wallowed and pitched, its topside invariably awash. Below decks were almost as wet. The galley was often flooded and Mawson complained of continually rolling out of his bunk onto a sodden floor. Early in the voyage, the main bilge pump was blocked by ashes and drinking water became scarce after one of the ship's tanks was contaminated with sea water.

Every inch of space on the *Aurora* was crammed with supplies. Difficult, dangerous or unwieldy cargo was consigned to the decks; so the expedition's sledge dogs shared their breezy accommodation with the Oregon radio masts, hut timbers, and as Mawson puts it 'six thousand gallons of benzine, kerosene and spirit, in tins which were none too strong, [that] we might well have been excused a lively anxiety'. Also stored outside, but out of reach of the dogs, were a couple of tonnes of butter and a few fowls. In total the 386 tonne *Aurora* sailed with 150 tonnes of stores, 380 tonnes of coal, 38 dogs, 24 officers and crew, plus 31 expeditioners.

On the stroke of eleven, an ear-cracking blast sounds from the *Shokalskiy*'s horn. Last kisses, handshakes and hugs are snatched as the band plays 'Now Is The Hour . . . '. But wait, there's a call on Rob Easther's mobile phone for Geoff! It's someone from the Heritage Commission wanting to discuss some minor point of detail—and not for the first time. The phone is hastily passed over the ship's railing to Geoff. His answer is a blunt 'can't talk now' followed by a hurried throw from his right arm. The mobile phone just manages to reach an anxious Rob Easther. The ship gradually edges away from the pier as streamers stretch then break. Two thousand five hundred kilometres to go.

The call to Geoff is typical of the bureaucratic inertia, shadowing some aspects of the project. The Antarctic Division themselves have been slow to act at times. As the body with statutory responsibility for Mawson's Huts they are obliged to consider our 'Initial Environmental Evaluation' or IEE: the Foundation's plans for minimising any environmental impact resulting from the expedition. As of today our IEE is yet to be approved by the Division's director, Rex Moncur, so technically, we could arrive in Antarctica without approval to set up camp, let alone work on the huts. Into this hiatus steps Dave Killick who files a story to this effect and so earns the momentary displeasure of both his AAP superiors and the Division's head.

Three more shuddering blasts on the horn and the last of the streamers breaks, each end still clutched by Joan and her partner, Judith. The ship turns through 180 degrees, then edges past the sheer sides of the cleverly decorated *Devil Cat* and on into the Derwent.

Before the ship is fully under way, and without time to really consider the significance of what's just happened, I help Dave Gillott and Geoff repack some of the hold. We're particularly worried about the beer, as much for the mess that would result in a big sea, as for the loss of a precious cargo. The hold could easily turn into a turbulent pond of stale beer. For reasons none of us quite understand, it's all been supplied in bottles.

Some cartons that had been left on deck are stacked in the hold, then wedged tightly to prevent movement. The work is hot and dirty but quickly done. Our first lunch at sea follows, then the pleasure of a well-earned snooze— unfortunately, but necessarily, cut short by a call to lifeboat drill.

The power of the Southern Ocean can never be underestimated. ANARE has suffered its share of damage on voyages to and from Antarctica, one of the most notable occurring in 1988 when three helicopters were smashed to pieces in the hold of the *Icebird* after cargo lashings failed.

Loss of vital equipment was of great concern to Mawson, particularly as so much of the *Aurora*'s cargo was carried on deck. Soon after leaving Hobart, the *Aurora* steamed into mountainous seas. Mawson remarked that:

> It seemed as if no power on earth could save at least part of the deck cargo. Would it be the indispensable hut timbers amidships, or would a storm break on the benzine aft and flood us with inflammable liquid and vapour.

The Mawson's Huts work party gathers for a brief meeting in the ship's bar late in the afternoon. Joan, Dave Killick and Rookie, all sport new close-cropped haircuts given to one another out on the cargo deck. Joan had been threatening this for some time, and I'm surprised, it looks better than I'd expected. Alan starts the meeting with a general vote of thanks for the efforts of the group, that's enabled us to get this far with such a good measure of harmony and cooperation. Of the rest of the meeting, I'm not too clear. That all too familiar sickly feeling comes back to me.

One beer after the group disperses, then the Southern Ocean takes over. As the ship clears South East Cape, I clear a path for my bunk. Dinner comes and goes without me stirring, but I do make a feeble attempt at both breakfast and lunch the next day.

Friday, 12 December

Night and day seem to merge. Time drifts by like the passing wake of the ship, as my head floats in one direction and my stomach in another. I never actually feel seasick when lying down; I just succumb to an overwhelming lassitude that can't be fought. If it weren't for the all-too-cheery messages gurgling at regular intervals over the ship's public address, I'd probably lose all sense of time.

The call to dinner finds me feeling a little more lively and I reach the dining table quite confident that the worst is over. The feeling, alas, is fleeting. As I mumble short and disinterested answers to a couple of passengers' solicitous enquiries, a cold sweat spreads across my forehead. I just manage to stay there though, stoically trying to think of anything but the motion of the ship. Then dinner arrives. I'll never learn, for, as usual, I'd been a bit overconfident with my order—one look at the large plate of spaghetti marinara and I'm gone, straight for my bunk, not even pausing to take off my shoes.

I'm not alone in my plight. There weren't many passengers at dinner and our group's numbers were also thin. Joan had been dreading the trip south and Dave Gillott vowed that we wouldn't see him until the *Shokalskiy* reached Commonwealth Bay. The crew, of course, are immune; so too the Southern Heritage staff. I suppose that's to be expected, as in their midst are some seasoned sailors. In the absence of Rodney Russ, founder of Southern Heritage Expeditions, the voyage is being led by Belinda Sawyer, a young New Zealander with a ship master's ticket and a wealth of marine experience gained all over the world. Also on board and working for Southern Heritage are Don McIntyre and his wife Margie. Don is a well-known yachtsman and veteran of the BOC Challenge single-handed race around the world. This voyage represents his fourth visit to Commonwealth Bay. Not that his second could exactly be described as a 'visit', as on that occasion, he and Margie stayed for a year. They sailed down at the end of 1994 in their yacht *Spirit of Sydney*, and, like Mawson, took all their needs with them, including a prefabricated hut.

The hut remains at Cape Denison and has been home for this last year to not two but one. A Sydney man, Alfred Winklmayr, negotiated with the McIntyres to spend the entire year by himself at Cape Denison, living in their hut. To my knowledge he may be the first person to endure Antarctica in absolute solitude for any extended period since the famous American explorer, Richard Byrd, spent four and a half months alone on the Ross Ice Shelf in the winter of 1934. Alfred is to return home on the *Shokalskiy*.

After helping to unload the ship, the McIntyres intend to stay at Cape Denison while our group works on Mawson's Huts, and return with us on the *Shokalskiy*. A party of friends and keen sailors aboard the *Spirit of Sydney* is due in Commonwealth Bay between Christmas and New Year. While in Antarctica, they'll join Don and

Margie on a trek to Madigan's Nunatak, the closest rock outcrop on the ice plateau to Cape Denison.

As well as being highly experienced sailors, Don and Belinda are also skilled inflatable boat drivers. They need to be. Together with Aaron, Rodney Russ's son, they'll be running all our freight ashore, as weather permits, in three four-metre boats. In the whole expedition, open-boat operations will probably be the most hazardous occupation.

Pinned in my bunk by seasickness, the thought of open-boat journeys in Antarctica is one that I'd best avoid. What was that saying about 'the devil and the deep blue sea'? Perhaps only now do I grasp the full hopelessness of such an impossible choice. Ten minutes after my hasty exit from the table, there's a knock on the cabin door. Margie's been checking on the non-starters and early finishers. As usual, a few minutes in a prone position is all I need. If not exactly well, I'm feeling hungry again. We both decide that I'd have a good chance of keeping dinner down, so Margie returns with my untouched spaghetti marinara. Over the next half hour or so I manage to slurp up most of it, with plate poised on my chest, and the whole lot washed down with lemonade.

Saturday, 13 December

As the seas abate, my sea legs develop. I seem fortunate, that despite feeling unwell for the first couple of days at sea, I invariably recover and am then only relegated to my bunk by very rough conditions. Like many, the longer I'm away from land, the better I seem to get.

I certainly tested and proved this personal maxim on my first big adventure at sea. Our vessel on the Heard Island Expedition was the well-known, 27-metre maxi-yacht, *Anaconda*. To reach the island, *Anaconda* had to negotiate 4,000 kilometres of stormy Southern Ocean south-west of Fremantle, in a gruelling passage that took nearly four weeks to complete. For the first couple of weeks, my stomach often complained, demanding to know why I'd been so foolish involving it in such an unnatural activity. But as we inched closer to the island, so I became inured to every lurch and roll. Two days from our destination, I was cook for the day, and with the assistance of just one other crew member, produced a roast dinner for nineteen, while outside, the waves crested ten metres and more. The meal was almost a disaster though. At one moment while we were cooking, the yacht rolled particularly hard. The oven door flew open and in a flash, a baking dish containing a large lump of beef shot across the galley floor, splashing fat everywhere as it just missed the legs of our very testy skipper.

Everyone is starting to relax and the passengers are easing into the cruising life. They come from a mixture of backgrounds: doctors, teachers, business people and the like, but as could be expected many share a long-held passion for Antarctica. Retired men, who couldn't convince their wives of the pleasures of Southern Ocean cruising, seem to be common among them, but there are also couples, and one family group with children aged 11 and 14. There's a marked absence of the 'been everywhere else' type; those who would be on board, merely to add Antarctica to their list of travel credits. All the passengers on this voyage seem attracted by the mystery and adventure of the place.

Now that I've reclaimed my appetite, I can enjoy the splendid meals prepared by the ship's two chefs, Phil and Leeson. For variety, freshness and style, their dishes would do credit to any restaurant. How they prepare such meals, day after day under extremely trying conditions, is beyond me—they make my *Anaconda* efforts seem quite insignificant. They're helped in the galley by several Russian women, one of whom has become a particular favourite among the passengers. Olga, like most of the crew, speaks very little English, but manages to charm all with her smile. Dinner orders are requested in summary by Olga as 'biff', 'chigun', fish or whatever single word describes for her each of the two choices of main course offered each night. Her attempts at English and ours at Russian are a source of great hilarity on both sides.

Despite the more comfortable ride in the last 24 hours, Dave Gillott has not ventured out of his cabin and Joan and Estelle, although up, continue to struggle. Only a few more days to go and all our planning will be put to the test. In the meantime, all except those unfortunates still afflicted by seasickness seem content to enjoy this pleasant hiatus. Maybe in a few weeks, when the Cape Denison blizzard is raging, we'll look back on the voyage with some fondness.

Our position at midnight is 53° 04' S 145° 47' E.

Sunday, 14 December

At around 4 o'clock this morning we crossed the Antarctic Convergence or Polar Front, the unseen, watery divide between Antarctic and temperate waters. At the convergence, temperatures drop by up to a couple of degrees Celsius, as cold, less salty water derived from melting sea ice, slides beneath the warmer surface water circulating southwards from the Atlantic, Indian and Pacific Oceans. It's at the Polar Front that the greater Antarctic begins.

Here, the Southern Ocean dons its habitual grey mantle, as fog shrouds the sea. The air chills rapidly and rain turns to sleet, then snow. But although sombre, dreary and cold, crossing the Polar Front is a memorable experience. The surrounding waters are rich in marine life, so seabirds become noticeably more abundant.

Giant petrels, and royal, wandering, black-browed and light-mantled sooty albatrosses all follow the ship, gliding effortlessly on wide, incisive wings. The sooty albatrosses, in particular, make balletic passes on either side of the *Shokalskiy*, dipping and weaving in pairs, crossing ahead of the bow then sweeping back, far to stern, almost teasing the ship to travel faster. Smaller fulmars and prions in flocks of about twenty birds, flit across the ocean's surface in a mercurial dance, their apparent fragility defying both distance and a merciless sea. But for me, the most delightful of all are the large flocks of Wilson's storm petrels. Breeding in the deep Southern Ocean and the Antarctic, it appears hardly probable that these tiny, butterfly-like birds, weighing no more than 50 grams, range into the northern Atlantic and Pacific Oceans on their pelagic migrations. About the size of a sparrow, the smallest of all seabirds literally skips across the surface of the most turbulent water, treating storms and gales with casual indifference. How they survive being blown to destruction, remains one of the great mysteries of the Southern Ocean.

At midnight the *Shokalskiy*'s position is 57° 29' S 144° 57' E.

Monday, 15 December

Average Southern Ocean conditions continue—cold, bleak and a grey eternity of fog-shrouded ocean. We've been lucky though, just managing to squeeze between two low-pressure systems. The day passes easily: chatting in the library, eating, then eating again after what seem to be ever-decreasing intervals, and of course sleeping.

Entertainment is plentiful for those that are interested, with regular slide shows, lectures and videos in the port-side dining room.

Shortly after lunch, Alan gives a talk about his days as station leader at Mawson in 1993. In an eventful and busy year, Alan oversaw the completion of the rebuilding of the station, and at the same time facilitated an active research program. 1993 also marked the end of an Australian Antarctic tradition when the last sledge dogs were removed from Mawson and the continent altogether.

Working dogs had been trialled at Heard Island in the 1950s and were used on the continent at Davis Station, as well as Mawson, until the mid 1960s. But by 1992, only 24 dogs and a few pups remained, all of them at Mawson. As a condition of the environmental agreement governing Antarctica—the so-called Madrid Protocol of 1991—all alien species (with the exception of humans) were to be excluded from the continent. As a result, the last remaining nations to keep sledge dogs (Australia, Britain and Argentina) all agreed to remove their teams by the summer of 1993–94. Most of the dogs were taken from Mawson in 1992, with our own Rookie responsible for their transport to an outdoor recreation centre at Ely, Minnesota. The journey was the subject of a documentary, *The Last Husky*, shot by yet another member of our current team, Mal Ludgate. Five old dogs remained at Mawson during Alan's year and were then brought to Australia and offered for adoption. And so ended a tradition that began with the first expeditions to explore Antarctica. Although the dogs' role as work animals had long been usurped by vehicles, their removal was hotly debated. As pets and a valued aid to morale, they are even now, greatly missed.

It appears to be a day for lectures, as I'm persuaded to give an afternoon talk on photography—'tricks of the trade' with wildlife and landscape. Not that my technique with wildlife is anything to emulate. A good friend, Morea, complains that next time she's with me when I'm photographing birds, she'll bring a pot of glue. In her opinion, the only way that birds could be persuaded to stay in one spot long enough for me to release the shutter, is if they're immobilised. Some glue slopped on their feet just might help! In other words, I'm far too slow.

Mal joins the session, and between us, I think we give some useful hints. In relation to Antarctica, our main piece of advice is that you can never take enough photographs. Film is cheap, the cost of returning is not. Lay your senses bare to the experience: allow yourself to be challenged by what you see and feel, but don't let the tunnel vision of a camera lens take control. Everyone seems satisfied, as we are also able to assist those more interested in f-stops and filters.

As the *Shokalskiy* pushes south, it's easy to forget that much is changing outside the snug confines of the ship. Forays on deck now require appropriate clothing, as the temperature is down to about zero and the wind quite blustery. I've been tempted to try painting from the ship as I did during my last voyage on the *Shokalskiy* but I suspect that I've let my opportunity slip. Nevertheless, I don't need convincing of the need for fresh air after the discussion on photography, so I head for the roof above the bridge deck. With its elevated 360° view, the top of the ship offers the best panorama of the passing ocean. Only from here can one start to appreciate the insignificance of our little floating sanctuary in the vastness and total indifference of the Southern Ocean. The damp and gloom confirm my thoughts about painting, but at least after fifteen minutes of wind on my face, I start to feel at least half alive.

At midnight the *Shokalskiy* lies at 61° 42' S 143° 55' E and the days are getting markedly longer.

Tuesday, 16 December

I sleep until 8 o'clock, something that's easy to do in this large steel cradle. Confinement, restricted air, slight seasickness and constant motion are a soporific combination. The perpetual daylight may also be starting to have an effect. I don't quite know when to sleep and when not, so when in doubt sleep anyway.

Today's great excitement is the sighting of our first iceberg, a heavily eroded tabular form with two distinct portions showing above the waves. A 'first iceberg sweep' was conducted among the passengers, where we all registered the time at which we thought a visual (not radar) sighting would be made. My guess came and went yesterday.

Only a few hundred metres long, the berg is nonetheless impressive. The distinct shape of tabular bergs derives from their having calved from the floating Antarctic ice shelves. One half still retains the flat-topped tabular profile, the other is more heavily eroded, with its slopes rising in sweeping parabolic curves to several spires of ice. The seas around are quite subdued but broad sheets of spray still erupt from its margins. Birds are everywhere, wheeling and gliding through the passage between the iceberg and the ship. On the flanks of the berg's eroded half, about a hundred or more dark specks, possibly cape pigeons, roost on the ice.

The sea picks up and is quite lumpy from mid afternoon on, with a south-easterly rising to 40 or 50 knots by evening. The fun for the day is a beach party in the bar. Quite a few come dressed in the height of tropical fashion. Were they hopeful of sporting such attire further south? Geoff appears resplendent in a Hawaiian shirt. Good silly fun as the ship lurches along through the storm-tossed Southern Ocean. As the evening progresses, the sea state worsens. Eventually the ship's course is altered 30 degrees and our ride smooths out a little. Everybody is anticipating pack ice in the next few hours.

Wednesday, 17 December

At 3 a.m. Belinda's all too cheery voice gives the call of 'pack ice' over the ship's public address. A couple of minutes later there's another more insistent call that finally draws me out of bed and up to the bridge. Many eager eyes are already there, staring out in wonder at one of Antarctic travel's greatest sights.

This must be my fifth encounter with the pack but my enthusiasm is undimmed. Each time it's like pushing through an infinitely variable mosaic, whose pattern changes with the time of day, season, sea and ice conditions, direction of travel and weather. While the thrill never diminishes, it's the first couple of encounters that retain the most vivid memories. My first experience of the pack goes back to 1984 on a round trip to Casey Station aboard the Antarctic supply ship *Icebird*. On the return voyage we negotiated the ice in the early hours of a January morning. Under a low sun, pack ice and icebergs shimmered with gold to the horizon and beyond.

On a more sombre day, with sullen skies weighing heavily in every direction, the *Shokalskiy* creeps toward the most favourable lead through the pack. We could be in the midst of a grand city where chaos was the inspiration for its layout. Aqua-edged slabs of ice, like ruined expanses of paving, are cut by gashes of open water; dismal bitumen-black paths disappear in futile mazes and blind alleyways. On either side, distant bergs flank our path. The faces of these imposing monuments in ice appear mostly bleached against the storm sky, but an occasional berg catches gold as a shallow slice of sunlight knifes through torn clouds.

It's easy to marvel at the ever-changing scene from the warmth of the bridge, but only by stepping outside is it possible to expose all the senses to the raw power of the experience. On the companionway aft of the bridge, the wind is more bitter than I had anticipated. A few minutes and some quick photographs are about all any of us can stand. Ice jostles along the sides of the hull making a dragging sound like wet sand, and intermittently there's a dull thud as the bow hits the next unbroken slab. At times the whole ship shudders as the ice resists its initial impact. When it finally cracks and the tension is relieved, the noise of ice falling against steel reverberates along the length of the hull. Against the random, changeable camouflage of sea and ice, snow petrels and Antarctic fulmars skip past in a 'now you see me, now you don't' dance, while loafing Adélie penguins are rudely evicted from their sea-ice rafts by the ship's onward push.

Large expanses of relatively open water follow. The captain and first mate peer at the horizon through binoculars or hunch over the radar screen, searching for the most promising lead in the next band of pack ice. The *Shokalskiy* is double hulled and ice-strengthened, but it's not an ice breaker. Like most ships that ply these waters, limited engine power means that steering a straight course through the pack is rarely possible. After about four hours of careful navigation snaking through the pack, we're back in open water for good. The next ice will be that of the continent itself. Away from the calming effect of the pack, twenty-five knots of south-east breeze makes the going quite lumpy.

By midday the *Shokalskiy* is creeping into Commonwealth Bay. We anchor well out from the abrupt ice cliffs that head almost the entire sweep of this enormous shallow bay and a cautious distance to the north-west of the Mackellar Islets, a small cluster of islands and rocks immediately off Cape Denison. These are dangerous and uncharted waters. From our distant anchorage, nothing can be seen of the locus of our pilgrimage—Mawson's Huts. The only discernible sign of human endeavour is the upright of the still defiant Memorial Cross, erected to the memory of Mertz and Ninnis.

We're all eager to get moving, so after lunch the decision is made to take the three inflatable boats to shore and reconnoitre the state of the huts, both those we'll live in and those we'll work on. A 20- to 25-knot katabatic wind is blowing off the polar plateau and the ship lies more than a kilometre and a half to the north-west of Boat Harbour. Tension is high as Mike, Mal and I clamber into a pitching and heaving boat at the bottom of the gangway. We're to be first ashore, so we can capture the drama of our party's first landing on Cape Denison. Mal clamps a vice-like grip on his highly sophisticated and precious Sony digital camera—the latest word in documentary gear—while Mike anxiously stumbles aboard, weighty sound gear slung round his neck.

All settled and hanging on, Belinda guns the throttle and we're straight out into the face of steep wind waves and freezing spray. At times we seem to be making little forward progress relative to the nearby Mackellar Islets but Belinda handles the boat with great skill. After a twenty-minute spine-jarring, tense and rough ride we clamber ashore at Boat Harbour. Despite a drenching, we're all fine, thankful that the cameras and sound gear are all well waterproofed in plastic bags or back packs. My finger tips are numb with cold, so I immediately remove both gloves and wring about a cupful of seawater from each. The other boats are fortunately right behind us. Dave Killick staggers out of the boat driven by Aaron Russ and announces breathlessly, 'I thought I was going to die'. Dave had been dreading the open boats, but despite the histrionics, his anxiety was genuine and, I think, quite justified.

The scene that greets our arrival is far more dramatic than the placid evening nearly twelve months ago when Geoff and I first stepped ashore at Boat Harbour. Instead of perfect calm, wind gusts down the valley from the south driving knee-high drift snow in smoky arabesques over the fractured lip of ice and out into the harbour. The locals are still here though. Lining the ice edge are about forty Weddell seals who slumber on totally oblivious of either the blustery wind or our arrival, and several small groups of Adélie penguins, gathered like well-wishers at a wharf.

Once everyone's ashore, we stow our emergency gear behind

some rocks: sleeping bags; a couple of tents; and enough food should we be unable to return to the ship for a while. Peeling off my wet-weather jacket, I bundle it up with the saturated gloves, which are hurriedly replaced with dry ones.

Through a stream of gossamer drift, Mawson's Hut looks totally different to last January. The place now has a bite to it that was lacking on that sublime summer's day. Although today's conditions are relatively mild for Cape Denison, I'm now more in awe at the skills of Mawson and his men in landing on this scrap of ice-bound rock and erecting such an enduring structure.

After a quick look at the hut, the three builders plus Geoff, Joan and Alan all head off on a tour of our potential accommodation options. Mike, Mal, Dave Killick and I tag along for the record. We first head over the ridge east of Boat Harbour to the Sørensen Hut, an ANARE field shelter erected in 1986. On the way we pass Gadget Hut, the McIntyres' neat and trim little structure in which they sheltered during their year at Cape Denison. The hut was named after the AAE dog 'Gadget', mother of 'Blizzard', the favourite pup.

Margie diverts us to meet Alfred Winklmayr—Austrian born, former ski instructor and real estate salesman from Sydney. Alfred is in the last throws of vacating Gadget Hut and preparing to board the *Shokalskiy*. He seems understandably reticent, although it's hard to judge from his covered and hooded face what he thinks of all the fuss and activity that has suddenly ended his year of isolation.

The Sørensen Hut comprises just one room, with a kitchen bench and a couple of tiered bunks, plus an entrance vestibule—the whole building measuring about five metres by three and a half. In the hold of the *Shokalskiy* we have sufficient panels to double its size. It was named for Arne Sørensen, captain of the much-loved and now sadly scuttled *Nella Dan*, a veteran Antarctic supply ship of the Danish Lauritzen Line. The Sørensen Hut was transported to Cape Denison aboard the *Nella Dan*. Typical of more recent field huts, the Sørensen is actually a refrigeration container with double-glazed windows, assembled from heavily insulated, prefabricated roof, wall and floor panels. Quickly assembled, the completed structure keeps the cold out and the heat in.

A quick and unanimous assessment by everyone determines that the Sørensen should be the focal point of our camp. It's hoped that with gentler winds, the panels for the extension might be landed on the ice edge directly below the hut, thus obviating the need to drag them overland from Boat Harbour.

Next, we retrace our route past Mawson's Hut and on to a small group of structures near the western edge of Boat Harbour. Set among rocks directly below Azimuth Hill and the Memorial Cross is the Granholm Hut, similar in construction to the Sørensen. Close by is an 'apple', a bright red dome-like hut just over two metres in diameter assembled from prefabricated fibreglass panels. Scattered to the north of the two huts are several rusting seatainers and a ragged line of old fuel drums. The Granholm dates from 1978 when an ANARE work party spent 40 days at Cape Denison attempting repairs to Mawson's Huts. It too is named for a Lauritzen master: Peter Granholm of the *Nella Dan*'s sister ship and transport for the 1978 group, *Thala Dan*. The Apple Hut was erected in 1985 by Project Blizzard.

While the Granholm and the Apple are convenient to Mawson's Huts, they are mostly surrounded by bare and very uneven rock that could be difficult to negotiate, especially when carrying loads or in high winds and blizzards. Our thoughts of living there dispelled, the decision to establish camp around the Sørensen is reconfirmed. It's also hoped that we may be able to relocate the Apple to near the Sørensen and so further consolidate our camp.

A small plywood 'cold' porch-cum-kitchen and pantry is tacked onto the northern end of the Granholm. It's a mess. The shelves ooze with stale and rotten food from Project Blizzard: bottles of grey olives, suspicious chutneys and limp packets of TVP, a substance that's every bit as appalling as its full name of 'textured vegetable protein' suggests. TVP remains all too familiar to me, a veteran of Project Blizzard's precursor, the 1983 Heard Island Expedition. If it were ever edible, it certainly isn't now. But there are also bars of chocolate and packets of biscuits that look none the worse for the passage of thirteen years.

About 6.30 p.m. we prepare to return to the *Shokalskiy*. The wind is still gusting strongly from the south and south-east, a typical Cape Denison gale. With the breeze behind us, the ride back to the ship is easy and we're there within five minutes. Getting out of the boat and onto the gangway is a different story though, as sharp wind-waves curl around the sides of the ship. The boats pitch violently. Joan is first to try getting out of our boat, but she seems glued to its pontoon, unable to make the great leap. The boat heaves up and down several times, then smashes into the bottom of the gangway. Suddenly she's away. I'm next but no more confident, making two aborted lunges at the elusive platform.

Each time I make a grab for the rail the boat drops and simultaneously slews away from the ship. Several waves break over the boat before I get my timing right. Then, in an instant I feel the reassuring solidity of the ship underfoot.

So is this what Antarctica is really like? After a calming meal, several drinks and a hot shower, my cabin and the prospect of a few hours' sleep beckon. But, as I will probably come to realise quickly while at Cape Denison, the weather makes all the decisions around here. By midnight the wind has dropped considerably and unloading begins in earnest.

Thursday, 18 December

The disassembly of the tightly wedged, organised jumble in the hold of the *Shokalskiy* is under Paul's steady direction with Alan, Dave Killick and myself either working in the hold or on deck. Others in our group are receiving the cargo ashore or crewing in the boats. Boxes, gas bottles, building materials, prefabricated panels for extending the Sørensen Hut, generators, sleds and tents all emerge from storage—hoisted by the ship's deck crane in cargo nets or suspended on heavy slings. Each load is then slung over the side to the waiting boats.

Awkward loads are no problem for Nikolai, the *Shokalskiy*'s bosun, as he deftly manoeuvres the crane back and forth while perched in its small cab. Wearing only thin greasy overalls in all but the coldest weather and forever sporting his distinctive headband, Nikolai looks like he's worked on ship's decks all of his life. Such is his experience that the lack of a common language is no problem. Hand signals to Nikolai work well: 'up', 'down' and 'stop' being the essentials.

The ship's hold is small, with half its capacity not directly accessible through the hatch. This means that some items must be dragged from its far corners before they can be lifted clear. In one such manoeuvre a pack of Baltic pine shifts and pins my left leg as I try to free the load from entanglement in a bundle of hydraulic lines. Out of sight, it's an anxious wait for what seems like minutes before the timber moves. Fortunately the crane takes up the strain on the sling and the load suddenly slews forward, away from my leg. Its full weight could have resulted in some bad bruising or worse.

In the perpetual light our normal rhythms are fast being countermanded, as rest, sleep and even hunger are staved off by the urgency of our task. With the hold half empty, we decide to pause for dinner, our first proper meal for the day. Although there's hardly a word of common understanding between Nikolai and ourselves, the camaraderie of the occasion and the sight of so much cargo gone from the ship causes him some satisfaction. As we turn for the warmth and comfort of the ship's interior, Nikolai's usually glum, weathered face suddenly cracks into a broad smile, exposing his two spectacular rows of gold-crowned teeth.

We now seem to have split from many of the passengers, like shift workers from the general workforce. As they end their day, we keep going—sleep can wait a while longer. Late in the evening we land below the Sørensen Hut and add to the depot left on the ice before dinner. It's slippery work from the landing to the depot over nodules of brash ice washed up and weathered smooth by the tide. The worst of the items is the stove which takes four of us to manhandle out of the boat and up onto the ice foot. Four saffron-clad figures struggling across Antarctica with a gas stove strikes me as being worthy of the absurd imagination of Spike Milligan or Peter Sellers.

Once again we work through the night with only the faintest of breezes picking up in the early hours of Friday. We finish work back at Boat Harbour, unloading the beer and other essentials. Despite the near calm, everything is tied down carefully, ropes interlaced over, under and through heavy objects which will act as deadmen. By 3 o'clock on Friday morning we're all back on board the *Shokalskiy* and ready for a few beers.

Friday, 19 December

Picture perfect, the way Commonwealth Bay is so rarely—like the best of beach-side Sydney and the Australian Alps rolled into one. Unloading begins after lunch. I crew for Don McIntyre and together we sling several loads of timber over the bow of the *Shokalskiy* straight into our waiting dinghy. In such gentle conditions, it all happens smoothly. All we have to do is position the boat under the descending load, manoeuvre it down so that its weight is evenly distributed, then release the sling. Don's hand signals to Nikolai are quick and precise and within minutes the boat's loaded with two bundles of timber. I can't help thinking what a difficult and dangerous task this would become if wind waves were curling around the ship's bow. Let's get the remaining loads ashore quickly!

Don guns the throttle and we're skimming to shore, an escort of penguins momentarily porpoising just ahead of the dinghy. As quickly as they appear, they're gone, swerving and gliding under the boat in effortless submarine flight. Within minutes we're at the ice foot in Boat Harbour, where we hurriedly break the strapping and plastic wrap on the bundles and pass individual sticks of timber to those waiting on shore. It's then back to the ship and the same again.

Drifting in the glassy shadow of the *Shokalskiy*, anticipating the next load to be slung over the side, the crystalline, deep-ultramarine water looks almost inviting, with the ice edge but a brisk swim away. In reality anybody unfortunate enough to fall in would only have minutes before death.

With load following load, the three boats scoot back and forth at a frenetic pace. Simultaneously, cargo is being ferried to Boat Harbour and to the ice edge directly below the Sørensen Hut. As more and more cargo leaves the ship our spirits rise higher. The task of unloading tons of materials and equipment is now within sight of completion, and the day that a few hours earlier we thought couldn't improve, just gets better and better.

After a few more trips crewing in Don's boat, I stay ashore at Boat Harbour and help drag the accumulated loads across the ice to a stockpile just above the highest tide crack, about 25 metres from the ice edge. The quad has now been landed and the unstoppable Dave Gillott puts it to work immediately, strapping large loads of Baltic pine boards across its carry-rack. With one person running along on each side to help stabilise the load, the quad quickly ferries all the timber up to the growing stack. The ship's passengers are marvellous. Most who are physically able, lend a hand for a while during their excursion ashore and by late afternoon several are still helping.

When the AAE began landing their stores at Boat Harbour on 8 January 1912, Mawson records that 'the day had been perfect, vibrant with summer and life'. But within hours, a gale was raging, making the task of retrieving the *Aurora*'s cutter and whaleboats perilous. Even though the *Aurora* was anchored in the lee of the ice cliffs, the waves coursing around her were 'at least four feet in height and when the wind increased to sixty-five and seventy miles per hour, their crests were cut off and the surface was hidden by a sheet of racing spindrift'. Two days passed before the gale abated and it was not until the morning of 11 January that unloading could begin afresh. The winds continued to hamper progress over the ensuing days but eventually all stores and equipment were ferried to the ice, including: 'twenty-three tons of coal briquettes, two complete living-huts, a magnetic observatory, . . . wireless equipment, including masts, and more than two thousand packages of general supplies'. In comparison, our unloading operations seem very modest, and peculiarly blessed.

Before the area around Mawson's Hut becomes a building site, cluttered with materials and equipment, I take time away from the unloading to secure a series of archival photographs of the building. These photographs will form a lasting record of the condition of Mawson's Hut and its surrounds immediately prior to our conservation work.

The virgin snow around the hut gives way underfoot with a pleasing crunch, although I feel it's almost a guilty pleasure defacing such a beautifully smooth and undulating expanse. In the stillness, distant voices, both human and avian, carry far across the valley with a crystalline clarity as extraordinary as the day itself. The temperature must be well above zero, and in the dazzling glare of the afternoon I'm able to strip down to just a tee-shirt and trousers, the bulky ventiles willingly, if only temporarily, discarded. Despite our feverish preoccupation with the tasks of the day, we're all so thankfully aware of what a blessing we have received. As we labour on, stopping only to gulp down some water or a soft drink, it's almost tempting to pull up a warm rock and lie back, savouring the splendours that surround us. But a lot more needs to be done and the boats are still running. Load upon load has just been dumped at the stockpile. It all needs to be sorted out and secured against the next gale.

By 8 o'clock we're all very tired, hungry and in need of a break, so it's back to the ship for a shower and farewell dinner. Once the meal is over, amid much hilarity and wine, Alan Grant presents AAP Mawson's Huts Foundation Work Party badges to the passengers and crew. The warmth and good humour of these people, most of whom we've come to know as friends, will be missed in the weeks ahead. The festivities continue in the bar with a number of passengers enjoying a free-flowing supply of beer, courtesy of the work party stocks. Alfred Winklmayr appears on board. As he sits quietly in the bar, I mistake his unfamiliar face for one of the less frequently seen Russians. I've not seen him without head covering and goggles before. Several of us slowly draw him into conversation. Before long, he's recounting with a mix of passion and longing, stories of his year alone at Cape Denison. He's so attached to the place and the ultimate solitude it affords, that he'd willingly pay to do it all again—now.

Saturday, 20 December

The weather early in the day is indifferent, with winds blowing at up to 20 knots. Boating activities are postponed until after lunch. Almost predictably, if that's ever possible in Commonwealth Bay, the wind eases around midday and by 3 o'clock the whole team is ashore with all personal gear, the fresh vegetables and other perishables. Furious activity begins immediately. The mountain of food boxes, drums, bags and additional tents are ferried up the hill by the quad and trailer or by hand. Ted and Paul fire up a generator and start drilling rock to house the bearer brackets for the soon-to-be relocated Apple Hut. By late evening the first of the bearers is bolted in place. Hard work is a perfect remedy to our situation as we're now totally on our own. The *Shokalskiy* has gone and our community is just thirteen on the edge of a huge and hostile wilderness. The realisation is sobering.

Our run of luck with the weather continues. For the first time since we arrived in Commonwealth Bay, the whole sky clouds over but the breeze thankfully remains light. By midnight we all feel we've done enough. Its time to knock off and savour a few beers, sprawled out on the rocks in front of the Sørensen. Joan and Estelle have the kitchen in operation and while the fare is comprised of what's readily to hand—smoked oysters on biscuits, soup and reconstituted freeze-dried 'beef and something'—it tastes wonderful. Our first 'home' cooked meal on the Antarctic continent.

And so to continue the run of firsts, it's to be our first night in tents. Geoff and I grab one of the long red nylon bags and shake a polar pyramid tent out onto the snow. The design of tents for Antarctic conditions hasn't varied much since the heroic era. The square or five-sided pyramidal shape has proven through long experience to be the most stable in high winds and was originally derived from Eskimo skin tents. The only fundamental difference lies in the materials: nylon fabric and aluminium poles having replaced canvas drill and bamboo. Like Mawson's tents, the frame and the cover are not separate, allowing quick erection in high winds.

Under the tutelage of Rookie, our most experienced Antarctic camper, we excavate and bench a square in the snow large enough to accommodate the fully spread tent. The tent poles are then unfolded and telescoped together and the two upwind legs spread and laid on the snow where they'll be stepped. Next, the main upwind guy-rope is used to pull the tent upright before the other two legs are spread into position. Finally all guys are stretched and staked into the snow and the tent's skirt, otherwise known as a flounce or valance, is spread out and snow heaped on top. Everyone helps, and soon all the tents are up.

The Sørensen camp is starting to look like an established settlement. Our tent is third down the hill in a line of six. Joan and Estelle are our immediate neighbours to the south and Mike and Mal are nearby on the downhill side. Rookie and Alan are furthest from the Sørensen, about 25 metres down the hill, while Paul and Dave Gillott's tent is closest to the hut. Ted scores a tent to himself. Rod and Dave Killick are possibly the unlucky ones, having opted for a smaller field tent, initially envisaged as a possible shelter for me when out and about. How long we stay in tents will depend on the relocation of the Apple and the extensions to the Sørensen. Joan christens our little street, 'Sørensen Rise'.

With all our clothes, sleeping bags and other paraphernalia inside the tent there looks to be barely enough room for us, let alone a couple of ANARE-issue camp stretchers. I decide to try the more direct method and bed down directly on the snow. That's the way I've done it snow camping in Australia. Admittedly those stints have only ever been for one or two nights at a stretch, so I think I'll be trying to squeeze a stretcher into the tent before too long. My bed is a piece of closed-cell foam, plus my own inflatable mat known as a 'Thermorest', and an ANARE sheepskin. I'm quite comfortable at first. Absolutely exhausted, I quickly fall asleep to the gentle flapping of the tent and the distant cackle of penguins. Inevitably it's not long before the wind gets up and the tent starts to flog. Although too tired to really care, the unfamiliar sound means that I wake intermittently and my hips get a chance to tell me of their displeasure at sleeping on snow.

Sunday, 21 December

About an hour after going to bed early on Sunday morning, I begin to regret that second beer. By nine o'clock I can't bear it any longer and lurch around the tent trying to don enough windproof clothing to make it to the bucket and back without freezing. The relief is blissful and the reclaimed sleeping bag even more so. I then lie in until just after midday and would have slept on if not for the unfamiliar sound of a generator. The day starts very late but we're all exhausted after the exertions of getting a foothold on this place.

Despite my tiredness, the generator makes me scramble, half guilty, for some clothes. There's obviously work to be done. Once again, I fight with my ventiles, then stumble up to the hut. Joan is welcoming, and directs me to hot water, a towel and an assortment of breakfast-cum-lunch offerings: porridge, and what she contemptuously refers to as 'mucilage' (muesli), sledging biscuits (known to all as 'sledgies'), spreads, fruit juice and coffee. It's still all chaotic and impromptu but under the circumstances the hut has become a haven.

Despite a katabatic wind that blows all day at a consistent 40 knots and above, some outdoor work is possible. Mal, Rod, Dave Killick and I excavate the snow bank immediately to the south of the Sørensen, then dig a series of holes in the ice to accept the steel pipe supports for the hut's extension. Meanwhile Ted and Paul continue work on framing the platform for the relocated Apple Hut. It's hard physical work in the wind and cold, but the prospect that we might soon be luxuriating in not one but two roomy huts is ample incentive.

Later in the afternoon, Geoff, Estelle, Alan and I head for Mawson's Hut for a first proper look with the conservation program in mind. Compared to the previous January, there's a lot less snow covering stretches of the hut's immediate surrounds. Rubbish discarded by the AAE lies all about, but is particularly concentrated over the exposed rocks on the northern side of the hut.

What may have been rubbish to the AAE is treasure to Estelle. One of her major tasks over forthcoming weeks will be to map and record in detail what she appropriately refers to as the 'artefact plume', downwind of the hut. Like any garbage tip, it says much of daily life. Shards of glass—multi-hued crystalline fragments in green, amber and amethyst—lie among old boots, offcuts of timber, ceramic electrical insulators, broken stove parts, penguin carcasses, lumps of coal, tufts of hemp, seal bones, rusted food cans and a miscellany of unidentifiable metalwork. Estelle is anxious that the artefact plume remain as undisturbed as possible, for there's evidence here that Mawson's party, short of ready resources, engaged in concerted recycling and adaptation, sorting refuse into distinct piles, handy for reuse when required. To Estelle, the context of an artefact is critical, as important as the artefact itself. An item out of place could lead to an erroneous conclusion.

For the diminutive Estelle, Cape Denison is familiar ground after her six-week long archaeological assessment of the huts and their surrounds in 1985. But even before any of the rest of us has spent a week ashore, we're in some admiration of her past adventures here. She's made of sterner stuff than might at first be apparent from her stature. One of her favourite tales from the more Spartan

'Blizzard' days is of her nightmare at sea aboard the *Dick Smith Explorer*, a tri-masted converted fishing boat that served as Project Blizzard's transport to and from Antarctica. Throwing up as she rode out her turn at iceberg watch in the midst of a force ten gale—all the while strapped to the mast—understandably remains as one of Estelle's most vivid memories of leaving Antarctica. Despite her inescapable seasickness, she's certainly enjoyed the comparative luxury of the *Shokalskiy* this time around.

The walk back to the Sørensen is one of simple grandeur. Crossing the broad snow-clad ridge above Gadget Hut, almost the entire landscape of Cape Denison is visible. On the shoulder of the next ridge to the east sits our camp; seven red pyramids and a silver cube, brilliant against the snow. On the trudge home over the coming weeks the sight of the Sørensen camp appearing over the ridge is sure to be a comfort. But further on past our tenuous refuge, the distant ice cliffs are a reminder, if one were ever needed, of the totally unyielding nature of this coastline.

I feel an added chill as the lengthening shadows flood deep blue cavities in the cliffs while jutting columns of ice like enormous, yawning rows of teeth, catch the last of the afternoon's sunlight. Over my left shoulder, the dozen or so Mackellar Islets sit close inshore, the most distant among them only about three kilometres off Penguin Knob. Their snow-laden squat domes remind me of frosted cakes, random morsels sliced from the frozen whole. Outside their cordon, Commonwealth Bay stretches away to an inkwell blue horizon, flecked white with jostling wind-waves.

By nine o'clock it's meal time and we all crowd into the Sørensen. Were there many more of us, some unfortunate individuals would be forced to eat in their tents. Despite the squeeze we're still quite courteous and considerate of each other. The humour of our circumstance is apparent to all. Because of the weather, the Sørensen extensions have not proceeded quickly enough, so we're in for a few more meals like tonight's, juggling plates while playing 'musical flour drums' with our seating arrangements. The evening inevitably ends in a few more drinks with the stayers apparently lasting until two or three in the morning. Our routine, if we do actually have one, has shifted to breakfast from eleven to one o'clock, lunch about four and dinner after nine.

Monday, 22 December

The tent flogs violently all night but probably through sheer exhaustion I'm able to sleep through most of it. I wake often though, and have many vivid dreams. Waking to daylight in the early hours of the morning is very disorienting. My first thoughts are clouded by anxiety, not knowing the time or for how long I've slept. After fumbling for my watch, it's with some relief that I sink back into my sleeping bag knowing that I haven't slept on into the afternoon. It's going to take some time to get used to the seamlessness of the Antarctic summer. Apart from the angle of the sun or the glories of the evening sky, the days have no markers, no beginning and no end. It just gets colder when the sun dips behind the plateau.

Rather than abating, the wind gathers in intensity as the day progresses. About mid-morning Dave Gillott and I go for water to Alga Lake, about a hundred metres to the south of the Sørensen. Alga Lake is one of a handful of small glacier-scoured tarns scattered across Cape Denison. Although the surface of this and the other lakes is permanently frozen, meltwater remains trapped beneath its uppermost crust of ice during summer and so we're provided with a convenient and ample supply. But it's not as simple as just cracking the surface of the ice with a crowbar. Standing in the middle of the frozen pond, ladling water into a twenty-litre plastic jerry can with a saucepan can be a cold, wet and frustrating business. Water sprays everywhere, scooped out of the saucepan by the wind. We try various techniques and orientations to the gale, but still receive a soaking to gloves, boots, and ventiles. The walk back to camp is wind assisted, but no less tiring than the uphill struggle as we each try desperately to control two heavy jerry cans while careering along before the blast.

Thankfully we're not intending to be here for winter when water gathering, like most domestic tasks, proves far more arduous. With no meltwater, solid ice is all that's available. In daily life at Mawson's Hut, gathering ice for eighteen men fell largely to Leslie Whetter, surgeon and assistant medical officer. In *The Home of the Blizzard*, Mawson recounts his struggles against the wind, procuring ice from the 'unfailing fountain-head of the glacier' noting that:

> Boxes in which Whetter used to carry ice for domestic requirements were as a rule short-lived. His problem was to fill the boxes without loosing hold of them, and the wind often gained the ascendancy before a sufficient ballast had been added.

By early afternoon, gusts of over sixty knots are ripping over the Sørensen. Even a trip to our stockpile of food to pick up a drum of flour for Joan is a trial. Going down is fine but the trudge back directly into the wind is energy sapping. Any movement outside becomes a deliberate test of strength and determination. Just getting in and out the door of the Sørensen becomes an effort. Because of the uneven terrain, entry is across a narrow verandah, right in the full blast of the gale and over a metre above the surrounding rocks. Once the struggle with the outward opening door has been won, it's then a matter of quickly throwing caution to the 'you know what' and jumping or gingerly climbing down while hanging on to a conveniently located guy wire.

Late in the day the maelstrom reaches its peak, at times surging to 80 knots and beyond. The sea in the shadow of the ice cliffs across the bay appears quite calm, yet away from their protective heights, the water seethes with a menace that we haven't seen so far. Above, the plateau is furred with drift turning pink in the evening sun, a softness of outline that belies the fury around us. Even the penguins appear to be having trouble. Bowled along by the wind, with feathers on the back of the neck blown flat, they seem not to have much choice as to their destination. I think if I see one flying past the window, it'll be time to worry. All about is chaos.

We expected winds like this, but nothing can prepare you for the experience. I've seen winds of this strength before, during my first visit to the Antarctic region in 1983. On Heard Island, arguably the most remote speck of land on the planet, wind speeds can reach 100 knots and more. Together with eleven others, I endured six weeks camped on one of the island's volcanic sand beaches. Violent winds often buffeted our stretch of coastline, which, on a couple of occasions blew with such strength that we were reduced to crawling on the sand. On Heard Island it would be calm in one spot, yet 50 metres away a sandstorm would be raging. Next minute, the wind would stop as quickly as it had sprung up, then roar in again on exactly the opposite tack. While the island is to me, more menacing and unpredictable than Cape Denison, its winds do not blow with anywhere near the same brutality or frigid, knife-edged monotony of speed and direction.

Photographing the storm proves difficult. Quick forays onto the verandah are about the best I can do. With arms propped defiantly on something solid like a gas bottle or generator, I try to steady the camera leaving just enough time for a few hasty shots before the wind chill becomes unbearable.

Throughout all the mayhem, Rookie has been working steadily, if not always quietly, getting the HF radio, fax machine, satellite dish and phone, computers and batteries installed and fully operational. We can now send and receive e-mails and faxes, and make satellite phone calls. Rookie will also conduct regular radio skeds with the *Shokalskiy* while she's at sea and with Casey, the closest Australian Antarctic station to Commonwealth Bay. Mawson, would no doubt be impressed by our near 21st-century technology as he was at the forefront of innovation for his own era.

The AAE was the first expedition to use wireless in Antarctica, but it took them more than a few days to establish communications with the outside world. Firstly, a party was landed at Macquarie Island on the *Aurora*'s southward voyage. Radio masts were erected and a relay station maintained on the island for the duration of the expedition. At Cape Denison it took from April to September 1912 even to erect the two enormous Oregon masts on which the radio aerials were strung, and it wasn't until February 1913 that two-way contact with Macquarie Island was made. What a difference 86 years makes!

Tuesday, 23 December

The wind persists, blowing at an unrelenting 35 to 45 knots for most of the morning. As nagging and tiresome as it is, the wind is fast becoming such a fundamental part of life that I can see in time that we will attempt more and more challenging outdoor tasks despite the conditions.

We get up in the morning and go to bed in the evening accompanied by the sound of the wind. In a short time it has taken over large portions of our consciousness. It's always in our ears; like nature's muzak, annoying and distracting, a seamless envelope of noise that smothers most other sounds. Our physical co-ordination is changing as we learn to 'windwalk': our bodies canted at strange angles to the blast or braced against the gusts. In most parts of the world wind is an intermittent element of the natural process, usually benign or beneficial, as everyday as sunrise and rain. A breeze blowing through the trees or in our faces is a sensual reminder of the fluidity of the air we breath. But midstream in a raging river of freezing wind and snow, the air is not measured in the unconscious rhythm of one's breathing or the gentle cooling of skin. Instead, there's a real anxiety of being overwhelmed, of drowning in its torrent. Of course our experience of the winds at Cape Denison, no matter how ferociously they might blow, is sure to pale beside that of the winter storms. In *The Home of the Blizzard* Mawson describes the combination of wind, darkness, and drift snow as one of 'suffocation' ten times worse than wind and cold alone. He writes:

> A plunge into the writhing storm-whirl stamps upon the senses an indelible and awful impression seldom equalled in the whole gamut of natural experience . . . We stumble and struggle through the Stygian gloom; the merciless blast—an incubus of vengeance—stabs, buffets and freezes; the stinging drift blinds and chokes.

Today is a significant moment if only for the symbolic event of removing the verandah door to Mawson's Hut. Dave Gillott does the honours. It comes away easily, regular use having ceased long ago. The steel strap hinges are broken and the door is secured only by a few nails. Once it's off, Dave fires up the electric chainsaw (after some difficulty) and makes easy work of what had looked like impenetrable ice. As Dave removes block after block from the passageway, Geoff, Mike, Mal, Estelle and I each intermittently form part of a chain to ferry blocks out of the hut to a dump a few metres away. To avoid trampling on historic material around the hut we shovel drift snow to form a ramp to the outer door.

On our return to the Sørensen we find that those remaining have brought the rest of the food and other supplies up from the waterside stockpile, running numerous sled loads behind the quad. What a useful little vehicle the quad's proving to be. It almost didn't come on the ship as there was some force of opinion in Hobart that on the rocky Cape Denison peninsula its usefulness might be limited.

With every chance of the wind intensifying yet again, Rod and I spend over an hour retying the remains of the stockpile. What's left are those items most likely to be blown away: the prefabricated sandwich panels for the Sørensen additions and the fibreglass sections for turning the Apple Hut into what the Antarctic Division cutely describe as a 'Melon'.

We finish work for the day at ten o'clock. In the perpetual daylight normal schedules seem hard to keep. Our work days are beginning late and ending late. Hours on the clock seem to lack all meaning. At first the late starts were the result of exhaustion following the unloading, now they have more to do with disorientation and the strong desire for relaxation after the rigours of the day.

Wednesday, 24 December

The wind is still blowing very strongly from the south to south-east so there's not much I can do outside the Sørensen. I take advantage of the forced confinement to complete a story I've been working on for the *Sydney Morning Herald.*

It's a frustrating business working on a lap-top computer in the crowded hut. The story takes longer than I had planned, what with charging and recharging batteries, or moving from one cramped corner to another as someone needs to find the onions, a pair of gloves or their camera. Then there's Rookie, who, justifiably nearing the limit of his patience, just wants to get to his own corner of the hut and continue setting up all his communications equipment. The room is filled with chatter and try as I might, I can't ignore David Killick when he puts on one of his comic turns.

By early evening I've had enough, my concentration completely gone. Joan is of a like mind. In common with most of the group, neither of us have washed any further down than our necks since the *Shokalskiy* departed. We also agree that it's bath time. The decision to have a bath is not taken lightly. At present the Granholm is the only washing facility open for business, so a round trip to the other side of Cape Denison is involved, as well as the rigmarole of the procedure itself. I think Joan detects a slight reticence in me, so she announces in her typical forthright way that she is past caring about modesty. So why should I care? I'm sure my need for a clean body will prevail.

Having battled the wind all the way across, I feel even more in need of the long awaited wash on arrival at the Granholm. We come armed with an extra saucepan to double our snow-melting capacity, changes of underwear, dirty washing, towels and ablutions gear.

Topped by two saucepans brimming with snow, the stove in the Granholm's kitchen cum porch cum toilet is soon sputtering along. The water seems to take forever to boil, but Joan and I aren't worried. The Granholm and its setting are far removed from the mild chaos currently prevailing at the Sørensen. The view north from the hut door is not the grand sweep of coastline seen from the Sørensen. This is a more intimate outlook, affording a chance to watch the passage of Cape Denison life at close quarters, and in near total comfort.

Several small colonies of Adélies are scattered along the rocky finger leading down from the hut and onto the western arm of Boat Harbour. The nearest group, one of about twenty adults, is quite close to the hut. As Joan and I gaze out the open door, watching the wind scuff around the Mackellars, a lone, curious penguin approaches almost to the hut, gives us a quick once-over then ambles off to its colony. Along the ice edge fringing the harbour, small groups of birds gather. They look almost like conversation groups, individuals gathering informally as they go about their daily business.

Water finally boiled, Joan and I divide the contents of the two saucepans into plastic basins. I then help by carrying hers to the Apple Hut next door and so solve our privacy dilemma. Luckily I don't spill a drop. At last prepared with a steaming basin of water, it's a relief and almost a forgotten pleasure to take all my clothes off. With the hut's door still ajar I stand naked in Antarctica, looking out on the penguin parade and Commonwealth Bay beyond. The water's too hot, so I bring the temperature down by adding a lump of snow, then step into my ankle-deep bath. First, a quick flick of water over the head, then a vigorous soap down, wash of the hair and rinse. Bliss. The remaining clean water is devoted to my underwear and socks.

With our damp underwear festooning the Granholm, Joan and I head back to the Sørensen, skin tingling and hair fluffy. I know the feeling won't last but the instantaneous effect on the morale makes all the inconvenience worthwhile. Stepping out across the flats below Mawson's Hut, I wonder how long the current nagging bout of katabatic winds is going to last. Are we to endure these conditions through all of our stay, and if so what will the group's morale be like towards the end? With more than a month to go, I can foresee our patience and stamina being sorely tested. Searching for distraction from these negative thoughts, I turn my attention to the changing landscape.

Hour by hour the snow continues to ablate and in the short time we've been ashore the whole look of the place has altered. Bare rock grows and in places the snow's quite slushy, making a delightful slurping sound underfoot. As more snow ablates, treading our path back and forth from Mawson's Hut could become quite treacherous, particularly in high winds. We've been provided with slip-on chains for our boots—they'll be fine for the ice but may be an encumbrance over rock. Walking anywhere around here requires that both be negotiated.

Back at camp, work proceeds in readiness for the Sørensen extensions and the arrival of the Apple Hut from Boat Harbour. Paul and Ted have almost completed the sub-floor framing for both buildings. Steel props are now level and plumb, frozen into the ice or grouted in holes drilled through rock. Tie-down anchors are similarly in place. Paul is an old hand at all this, a veteran of four winters building in Antarctica and almost as many years in the frozen wilds of northern Canada and Alaska. A few months before our departure, Paul was honoured with the Antarctic Medal for outstanding service to Australia's Antarctic program. Usually of a somewhat laconic nature, easy-going Paul has seen it all before. In his late forties, what he might lack in youth is compensated for by the experience he brings to the task at hand. His years of experience show in the way he moves around the site, slinging heavy loads on one shoulder with ease, or deftly handling his tools. He maintains his agility despite walking with a slight stiffness to his gait, the result of a serious fall from a timber bridge while working in Canada. The obvious question to Paul is why he continues to be drawn back to such cold and remote places. His answer is that 'cold places pay well', but we all know that's but a small part of the answer.

The preliminary work complete, all that's needed now is a still day to allow us to stand the prefabricated panels on the Sørensen extension. At about 9.30 everyone's back from Mawson's Hut and we all cram into the original Sørensen for dinner. Alan briefs us on the state of our fuel supplies as there's general concern as to whether our stocks will last.

Unbeknown to Alan, a 44-gallon drum of petrol intended for our use had been loaded aboard the ship in Hobart. In the rush to unload, it had been assumed as belonging to the McIntyres and so remained on board for them to deal with. The McIntyres thought the fuel was ours and also left it alone. Consequently, we'll be reliant for most of our supplies on a stockpile of old ANARE drums at Boat Harbour. Dave Gillott reports that there's only about three hundred and fifty litres left in the drums, with the strong possibility that some of the fuel is contaminated. The conclusion is reached that if we manage our stocks carefully, we should just have enough. This will involve hooking up batteries on charge to one of the generators at the same time as kitchen appliances or power tools or a combination of all three. Communications via fax or e-mail may have to be restricted. There are a few tense faces in the room, none more so than Mike, who seems to spend nearly every moment when he's inside hunched over a lap-top computer, churning out e-mail messages.

Adding to the general anxiety we learn that the *Shokalskiy* hit a rock somewhere out in the uncharted waters of Commonwealth Bay while returning from Dumont d'Urville to disembark the McIntyres. It's a reminder that not only small boat operations are hazardous around here. The *Shokalskiy* is now caught behind the pack ice that's rafting up ahead of the incessant katabatic winds. If the weather permits it may return tomorrow and hopefully boost our fuel supplies. Our day ends with lots of silliness in the early hours of Christmas morning.

Thursday, 25 December

Most of us are up early to see if the *Shokalskiy* has made it back into Commonwealth Bay for Christmas, but as it has blown 40 to 50 knots all night we don't hold any real hope. We learn from Rookie that the ship had come close in the early hours of the morning, however the winds put paid to any possibility of launching the boats. The *Shokalskiy* has travelled as far east as 147° looking for a lead through the pack ice, then back past Commonwealth Bay towards the French station at Dumont d'Urville, 100 kilometres to our west. We hope they'll find a way north soon.

The chance of spending this special day with our friends from the ship, not to mention the whiff of a hot shower, is gone with the wind.

Around midday the wind eases back to about twenty-five knots and the temperature climbs to a very pleasant four degrees Celsius. One surprising aspect of the early part of our stay has been the relatively mild temperatures—often just above zero. However, they are not much more than a psychological comfort since with a temperature of zero and a 50-knot 'breeze' the wind chill is still a miserable −18°C!

Over the past few days I've become quite anxious about the scarcity of good painting days, so this Christmas afternoon is an opportunity too good to miss. In my early travels around Cape Denison I'd thought that the point just to the north-west of the Sørensen, with its abrupt north-facing slopes, might offer some good shelter.

With great anticipation, I cross the snow below our camp then amble along a steadily rising ramp of rock to a pavement of broad, rough steps. The highest step is sufficient to block the breeze, so with glee I sprawl on a wide, sun-warmed slab, ready for my first real chance of painting in Antarctica since visiting Casey Station in December 1984. Although I made an attempt on my brief excursion to Cape Denison last January, that 4 a.m. effort ended in frustration, as my paints, wet paper, water and brushes all froze.

Snow petrels wheel in expansive loops, spearing into the wind then banking away in easy passes, dazzling white darts against an indigo sea. About ten metres below my vantage point, an occasional Adélie saunters across the last vestiges of the winter seas still frozen fast to the shore, while small flotillas of penguins cruise just below the ice edge. They duck and dive repeatedly, braying to their cohorts, both in the water and out, barking a sort of 'come on in, the water's fine'. Adélies always seem anxious around the shore, swimming close in, or when marshalled on the ice. Waiting to go to sea, they shuffle about, procrastinating and squabbling constantly, all the while slowly edging closer to the brink, before a single, brave or foolhardy bird dives in. The rest follow quickly. Their hesitancy is justified, as Adélie penguins are important fare for leopard seals, lone but lethal predators of the ice edge.

During winter, the sea ice would have stretched out into Commonwealth Bay, and at times of calm, all the way to the horizon. But because of the incessant winds, the ice off this coast is nowhere near as thick or enduring as that encountered elsewhere around the continent. Newly frozen expanses don't get much of a chance to consolidate, being constantly torn apart and blown northwards by the force of the offshore gales.

Away in my thoughts, I'm suddenly puzzled by an alien sound echoing from the rocks on the other side of the ridge. The pulsing noise intensifies, reverberating from rocky ridge to glacier and back. A helicopter roars over, circles, then lands on the snow about a hundred metres from the Sørensen. Four brightly clad figures clamber out. They're met shortly after by Alan Grant and the McIntyres; Alan having raced back from Mawson's Hut. Our visitors could only be from *L'Astrolabe*, the French supply ship that services Dumont d'Urville. I turn seawards and find the tiny outline of a ship riding far out in the bay.

As our visitors are close by, curiosity gets the better of me, so I temporarily suspend work and wander over. In the course of conversation, Alan mentions that we're short of fuel. Without hesitation, they offer us as much aviation kerosene or diesel as we could want. The trouble is, we need petrol and that's just what they don't have. The thought is greatly appreciated and amplified by the offer of some time aloft with cameras for Mal and myself in the coming days. *L'Astrolabe* plans to be off our stretch of coast for some time.

The painting complete, I feel greatly relieved to have made a start. If I can produce an average of at least one work a day for the remainder of our stay, I'll be very pleased with myself. It's not the quantity so much as the accumulation of a satisfying collection of work that's important. Each painting, even if sketchy or just the bones of an idea, will provide a small but potent reminder of my first response to the place. When I'm back in my studio, this assortment of quite modest scraps of paper will do more than any photograph in returning me to the landscape that inspired them.

I return briefly to the Sørensen to drop off my gear before heading to Mawson's Hut. Joan's in full production on the Christmas dinner. She's also in a state of high dudgeon over our French visitors. As it landed on the snow near the Sørensen, the helicopter also came to rest within a hundred metres of our closest neighbours, a colony of Adélies; or her 'chooks' as Joan likes to think of them. Penguins scattered in all directions, abandoning eggs and newborn chicks, trampling neighbours and their progeny in a brief attack of communal panic. Once the noise and the down draft from the rotors subsided, order was quickly restored, but disturbances like these leave the chicks and eggs vulnerable to attack from ever-vigilant skuas. Also, any increase in stress is bound to further skew the odds in the struggle for survival through breeding success. In the course of our daily activities around the Sørensen, we've been careful not to go any closer than necessary to nearby colonies. The thought of being the cause of the loss of an egg or a chick is enough of an incentive to keep well clear.

Joan's indignation is perhaps coloured by past actions on the part of the French. At Dumont d'Urville, the resident scientists, national servicemen and support staff are greatly outnumbered each summer by tens of thousands of Adélie penguins, nesting in the immediate ice-free surrounds. Back in the early 1990s, the French earned the ire of conservationists by flattening several offshore islets in the construction of an all-weather earth runway intended to ease reliance on shipping for the resupply and servicing of the station. The resident penguins did not readily accept their eviction notices and returned time and time again to the vicinity of their bulldozed nest sites. Nature had the last say though. Part way through construction, a massive fragment of ice calved from a nearby glacier tongue, sweeping the partly constructed runway into the Dumont d'Urville Sea. Construction has not recommenced.

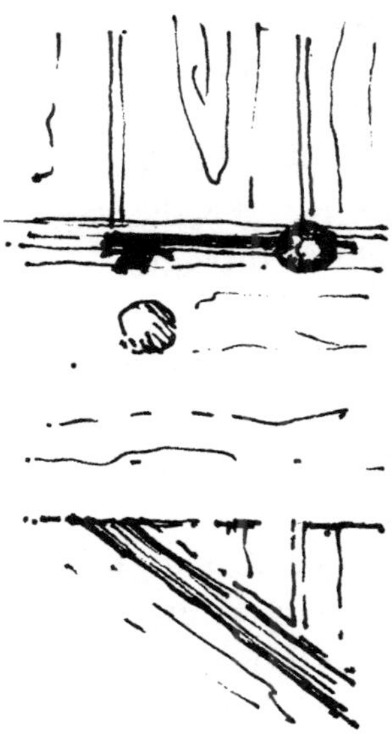

Over at Mawson's Hut, the tunnel is being pushed on into the workshop. I take time to photograph the progress now that the first of two removable baffles installed by Project Blizzard has been reached. To everyone's relief and amazement, the tunnel beyond has remained reasonably free of snow for the past thirteen years. As the afternoon sun streams through the tunnel, ice melts from the back of the porch door, revealing our first 'find' inside the hut. We're all amused by the sight of a key nailed to the ledge of the door. Mawson's Hut must be, with all its natural defences and remoteness, the most secure building in the world. For us, this wry gesture is a first glimpse of the lighter side of Mawson and his men.

Knocking off for the day, I hitch a ride with Alan to the

Sørensen, seated on the back of the quad. This is not an activity that ANARE might officially sanction should we be on an Antarctic station. Rod's not too happy about such behaviour either, but with Alan driving cautiously, it's fun, and I think quite safe—an exhilarating end to the outdoor part of an unforgettable Christmas Day.

Back at the Sørensen, Joan has concocted a miraculous Christmas dinner. As well she's decorated the hut and herself, sporting an outrageously overblown pair of earings that would look more at home on a tree than dangling from her ears. We learn that they actually belong to Estelle, who, not to be outdone, wears 'baby Jesus' earrings that are purported to glow in the dark.

All thirteen of our party plus the McIntyres cram into the hut, some sitting on flour drums, some standing and a couple half prone on the lower bunk. We feast on roast lamb and vegetables, followed by a tinned plum pudding and custard. Possibly from the ANARE visit of 1978 or from Project Blizzard in the mid 1980s, the pudding is very acceptable despite its age. The meal is well complemented by beer, wine and French champagne, courtesy of our sponsors. Cameras flash from every angle and Mike and Mal (instantly dubbed the 'Leyland Brothers' by Margie) try to film the festivities.

In the hubbub of hut banter and the growing camaraderie of our little group, our isolation from family and friends on this Christmas Day is almost forgotten. Outside, the evening gale increases in force and menace, turning Commonwealth Bay into a wild, chaotic sheet of spray and white water. As the wind quickens further, the Mackellars drift from view shrouded in a vaporous spume. To the east, the ice cliffs smoke with gale-driven drift and as the sun edges lower, airborne spangles of ice refract the light, drawing rainbow curtains across the sky.

We're all thankful for our little haven from the maelstrom. The overcrowding only seems to add to the occasion. We receive greetings from home via Alan from Rob Easther and from David Jensen and Les Murphy at AAP. There's a queue for the phone, but we all seem to get through to our nearest and dearest, just audible over the din. Like most, my conversation with the family is garbled and short.

Christmas wouldn't be complete without presents, and this most unusual of days is no exception. From AAP there's a package for each of us that includes a small 'point and shoot' camera. It's much appreciated by all. Even though I already have five cameras in my arsenal, the new one will be perfect for 'happy snaps' and a few prints to hand around at home.

Three members of our party went to varying degrees of trouble to organise presents for their fellows. Joan, concerned for the nocturnal comfort of the males in the group, gives each of us a hot water bottle, to be employed 'as we see fit'. The dual purpose implied in this gift causes much hilarity. Estelle's presents are intended to be totally useless. In keeping with her predilection for kitsch, bad taste and the useless, everyone receives a large, brightly coloured, plastic bottle stopper.

Geoff, on the other hand, has gone to a great deal of trouble with his presents and in so doing takes kitsch to new heights. As his cabin mate on the voyage down, I watched him labour for several days in less than ideal conditions, fashioning thirteen tiny but identical cut-outs of Mawson's Hut, that, with base and cover added, turned into Mawson's Hut 'snow domes'. We're all greatly impressed. I think Geoff considered his efforts worthwhile, judging by the uproar. I'll treasure mine and I know the others will too. A marvellous evening ends at about 1.30 a.m. with the inevitable confrontation with the elements.

Friday, 26 December

The day starts like all others with the struggle to get dressed. Most tasks in Antarctica take more time than one would expect—dressing is no exception to this maxim.

Several forces are at work first thing in the morning. Firstly, there's the thrashing tent, amplifying the awfulness of the weather to the point where it seems foolhardy to venture outside. Not being able to see out doesn't help, and then of course there's the pressing need for a pee. Getting going is essentially a conflict between the inertia of a warm sleeping bag and the anxiety brought on by a full bladder.

There's just enough room for Geoff and me in the tent, but with all our gear forming a mountainous divide between the beds, only one can stand at a time. The first to stir wriggles out of his sleeping bag, pulls on several layers of clothes, fights with ventile overalls and jacket, then boots, and eventually leaves before the other takes his increasingly urgent turn. I'm not sure whether it's preferable to pull your clothes on lying down, or risk falling onto one's tent-mate by conducting the struggle standing. Either way it's frustrating. Even crawling out of the entrance chute is a challenge.

When negotiating the two layers of this writhing nylon snake, the trick is not to become entangled in drawcords or toggles lest the day might start face first in the snow.

A 50-knot blast greets my first tentative engagement with the outside world. Free of the tent but squinting into the glare of sun on sea and snow, I struggle to my feet, then hunched against the sudden blast of cold, make the short dash to the Sørensen for a warm wash, hot coffee, breakfast and sanctuary. After a slow morning coming to after Christmas, several of us head for Mawson's Hut and more tunnelling.

Ice excavation has so far opened a passage through the verandah, dog legs to the right and left through two doors, and then on into the workshop. Ted, or 'Chainsaw' Ted as he's quickly come to be known, has taken over from Dave Gillott as our champion chainsaw wielder. With what we're coming to appreciate as our most valuable power tool, Ted deftly slices through everything from hard, near blue ice, to soft snow. He produces a steady supply of easily manageable blocks which Geoff, Estelle, Mal, Mike and I shuffle back along the tunnel to the snow ramp at the outside door. After a couple of hours, Ted reaches a second timber baffle installed under the central valley gutter by Project Blizzard in order to minimise snow ingress to the living section of the hut. We all stand aside, giving Geoff the honour of finally breaking through. But it's something of a false dawn as he pulls the boards away, revealing more ice and snow beyond.

As the tunnelling party develops the art of 'chainsaw archaeology' inside the hut, Alan, Paul and the two Daves erect the 'Weatherhaven', a large multicoloured tent that we plan to use for shelter and storage close to Mawson's Hut. Christened the 'Big Banana', it looks as though it might be more at home on a beach. It's a battle getting the thing up in the strong wind, and who knows how long it'll stay up? To help hold the Weatherhaven down we gather large stones from the moraine and add them as deadmen to the snow and ice piled on top of the tent's valance. Tools, generators and heavy boxes of hardware are moved inside for safe storage, further anchoring the Weatherhaven to the glacier.

The walk home is never a chore. It allows time to contemplate the landscape in a relaxed and meditative way. This is not a static place; changes are wrought not just daily but by the hour. Down from the Transit Hut, the eastern arm of Boat Harbour is iced over at present. In sunny conditions its surface becomes slick and shiny, streaked with aqua-blue melt pools. As the melt advances, slabs of sea ice will break away from the harbour's margins, their tenuous anchor cut by the combined action of sun, wind and tide.

Usually, about twenty Weddell seals, many of them recently weaned pups, bask on the sea ice. Weddells are an exclusively Antarctic seal, and while they're seen in the pack ice, they mainly frequent the inshore expanses of fast ice around the margins of the continent. When the sea freezes over in winter, Weddell seals maintain seasonally permanent breathing holes, keeping them open by sawing through the ice with their teeth. There were a couple of breathing holes, complete with resident seal, still in existence when we first landed at Boat Harbour. I was nearly the source of much hilarity, just avoiding stepping backwards into one.

Most of us arrive back at the Sørensen about nine o'clock to the smell of freshly baked bread and vegetable lasagne, but no Joan. She and Rookie are guests of the McIntyres.

Saturday, 27 December

The wind seems to have become a permanent part of our lives. It has blown furiously all night.

After the usual protracted breakfast and diary session, I decide to paint, no matter what the conditions outside. The nearest available shelter is only about twenty-five metres away, the combined influence of the hut and a large pile of boulders providing just enough protection. A couple of penguin carcasses act as my mute companions but I can't be fussy about what I sit in or near. I quickly unpack my gear and settle to work, anxious to make the most of the time before the cold defeats me.

The surrounding boulders warm to the December sun, and in shallow depressions and cracks, moisture won from melting ice allows cushions of green moss to grow. Small flecks of yellow and orange lichen dot the boulders like careless dribbles of paint, while other forms clump in dark hairy growths or warty thickenings. This is about as much plant life as the rigours of the Antarctic climate allow. Only two native vascular plants are found in Antarctica, both occurring on the more temperate Antarctic Peninsula. Over the remainder of ice-free Antarctica, only algae, mosses, lichen and one recorded species of liverwort are found. The lichens are the most diverse group, superbly adapted to the cold, dry and seasonally dark conditions.

Settled out of the wind, my painting proceeds quite well but after a couple of hours I retreat to the hut, a welcome mug of coffee and lunch. I'm pleased with the work, the first effort at my usual full size—about 50 by 35 centimetres. Provided there's shelter, I now feel more confident that I'll be able to paint outside in all but the more extreme conditions.

After lunch I head for Mawson's Hut. Meeting Geoff in the Weatherhaven, he tells me excitedly that during the morning the ice barring entry to the living section of the building had been breached. Eager to see for myself, I head immediately for the hut. A slight draft and the dank, musty smell of damp leather, paper and wood wafts back down the newly excavated passage.

It's dark at the end of the tunnel, the last smudge of light fading in the workshop. Donning a head torch, I'm dazzled momentarily by the bounce of light on snow but just ahead, a ragged opening, about a metre in diameter and a metre and a half off the floor of the tunnel, punches on into darkness. Barely large enough to crawl through, the opening drops a metre or more into a crystalline grotto encrusted with hoarfrost, stalactites and stalagmites of clear ice and fancifully sculpted mounds of soft, fluffy snow.

Almost as if the original residents had only just made a hasty exit, shelves and bunks are still scattered with everyday items; stacks of newspapers, books, tobacco and flour tins, broken scissors, medicine bottles, food cans and buttons. The people I had read about and imagined as they laughed, schemed and planned, worked and argued in this space suddenly become real. There, close on my right, is Frank Hurley's bunk (taken over by Arthur Hodgeman during the second year) inscribed in black paint with the initials and nicknames of both men and the dates; 'J.F.H. [Hoyle] 1912 and A.J.H. [Uncle] 1913' Convenient to Hurley's bunk, the door marked 'darkroom' stands wedged ajar by a large cornice of ice.

Directly behind the main wall of shelves in the living section is Mawson's room. Like the rest of the hut it lies in disarray, its contents in tatters. Encrusted with grotesque frozen chandeliers, a low ceiling of ice hangs ominously over the entire space and a lone chair, its back missing, sits askew near Mawson's bunk. Nothing much remains on his shelves: a few tins; a Petri dish and other fragments of scientific equipment; scraps of cloth; pages from a nautical almanac; a reel of cotton; a button and what look like the remains of an apple. Decorating the wall above the bunk, mounted on a square of grey paper and covered in spangles of frost, a single art print remains in place. I think it's by the French rococo painter, Fragonard. It depicts a woman in a mythical garden seated on a swing and dressed in ornate and voluminous pink skirts. Her suitor or lover reclines at her feet. Another print lies ripped in half on an empty shelf that once housed Mawson's library. The torn picture shows a satyr perched in a tree serenading his love on the pipes. The nymph sits naked in the grass at his feet, dreamily plaiting a daisy chain.

Reports of this print from Project Blizzard's investigations suggested that there was something pornographic or at least risqué about the picture. To me, it just seems a sentimental rendering of a classical tale, and nowhere near as titillating as the work of that most sensational Australian artist of the era, Norman Lindsay. To conclude that the picture was somehow prurient betrays an ignorance of art history.

The print's presence in Mawson's private domain was somehow thought to have rendered him less heroic. Here was evidence that he was subject to earthly pleasures and was not just the ice-honed demigod explorer. I've even heard mention that the print's current state—torn in halves—is possibly a disapproving comment from someone unknown who's visited the hut since 1913, removed it from the wall and vandalised it, in an effort to keep the Mawson myth pure. Was this meddlesome person unknown, Mawson himself on his return visit of 1931? Surely not?

I feel that all this flimsy supposition stretches a point. To look to these small scraps of paper for evidence of the human face of the man fails to take into account the bias that the myth-making of history leaves in its wake. That Mawson was a man, with ordinary human emotions and passions has been ignored in the creation of the hero. Inevitably, the historical legacies of men like Mawson suffer because of the strength of their actions and the beadth of their achievements. That they were also lovers, husbands and fathers is often forgotten. Anyway, I think with these pictures he was simply adding a touch of softness. They're all images of a faraway world, one of fantasy and imagination, that would've made the long winter days just a little more bearable.

All that remains of the bunk where Mawson recuperated from his sledging ordeal is its timber frame and a few loose slats. On one of the boards, obviously a remnant from a packing case, the name 'SY *Aurora*' is stencilled. At the head of the bunk, a bedraggled pillow droops over the side and a grubby candle looks set to roll between the slats and onto the frozen floor. Mawson, in common with other leaders of his era, maintained a hierarchical style of command. Reflecting this hierarchy, Mawson had his own cubicle while the other men were all in together, their tiered bunks distributed around the outer walls of the hut. His room had no door and was quite small. But it afforded Mawson some privacy where the others had none, and it reinforced his status among the men.

Nearly every surface in the hut is covered with sparkling frost crystals. Adjacent to the darkroom door a corner of the stove protrudes from the snow like some jewel-studded piece of priceless furniture. In places where the air has been still for years, exquisite gardens of ice, fragile needles of hoarfrost, grow in all directions.

After an hour or so taking some preliminary photographs, I emerge from the darkness to a balmy, gentle evening. As I amble over to the Granholm to see what's doing, the peace is disturbed by the clatter of a helicopter somewhere to the west, possibly towards Cape Hunter. It's a great day for flying, so the French are back. Over near the Granholm work has been proceeding apace in preparation for relocation of the Apple Hut to the Sørensen camp. This will be the first stage of a total clean up of the recent buildings and rubbish within the historic precinct of Mawson's Huts. If time and the weather permits, we will also move the Granholm and prepare the rusting shipping containers and empty fuel drums for removal by helicopter should the Antarctic Division's supply ship *Aurora Australis* be in the vicinity of Commonwealth Bay as planned next March.

Dave Gillott and Ted have readied the Apple for its journey, releasing the wire tie-downs and screwing the floor to its three support bearers. Ingenious and resourceful as ever, Dave has also made a sled, its ski nose cut from an old 44-gallon drum using only a hacksaw. With all hands helping, we lift the hut by its bearers and gingerly carry it across the rocks to the snow and the waiting sled. From there, it'll be dragged most of the way behind the quad. With Dave coordinating our efforts and barking orders to 'lift, carry and rest', this potentially hazardous manoeuvre is completed without incident.

Once the Apple has been firmly lashed to the sled, our load is coaxed into motion with Alan gunning the throttle on the quad. Coaxing also involves the combined exertions of eight middle-aged, or near middle-aged men, plus the youthful David Killick, pushing and pulling at just the right time. That's the theory anyway. It takes a few attempts, with much heaving from behind and straining on the tow ropes up front, before the sled begins to move. From then on assistance is only required on the steeper pitches—the ever-capable quad does the rest. Once over the rise and on the downhill slide to Alga Lake, our precious cargo requires restraint rather than coaxing for fear the sled might race out of control.

The ridiculousness of a near-spherical, red plastic lollipop scooting across the snow towards the polar plateau, while being chased by a mob of breathless, stumbling geriatrics doesn't escape any of us. It probably looks ridiculous to the French too. As we chase the Apple uphill, the chopper makes several low passes overhead, on the lookout for their geologists and ornithologists out roaming the hills of Cape Denison for most of the afternoon. We all think it would've been so much easier to sling the Apple under the helicopter, but the French never offered. Anyway, the job's almost done and there's growing pride among us in anticipation that we might actually win the battle. Meanwhile, as all the action unfolds, Mal, as well as pushing like the best, relishes the moment and films the entire circus.

At the head of Alga Lake we're back to the exhausting 'litter method', as six of us stumble over the rocks to the waiting platform just below the Sørensen. Those with the short ends of the bearers end up taking the heaviest back-wrenching part of the load, so every few metres we put the hut down and rotate through one position. As ever, this manoeuvre is orchestrated by Dave Gillott, who lets slip as we dance falteringly over the boulders, that an Apple weighs over 250 kilos. As if we need reminding!

With our latest addition to the Sørensen camp finally eased into place, Ted and Paul set about immediately tying the hut down to the anchors fixed into the surrounding rock. As the rest of us bundle into the Sørensen for a much-deserved beer and dinner, Joan remarks that the sight of the Apple appearing over the rise, man-handled by a team of 'slaves' was worthy of Cecil B. de Mille. And she's curious to know who the great ruler was inside our bright red palanquin?

The languid calm continues. Its pleasures are too rare to let pass, so dinner over, I head off with camera and tripod and wander the night for the first time. There's still a collar of sea ice extending out from the rocks below camp so I head down to investigate. Adélies loaf on its widest part extending about ten metres out into the bay, untroubled by a gentle shore break that occasionally licks over the lip of ice and spills across the platform's outer margins. Rounded nodules of brash ice washed up by the tide are scattered over its crusty surface like an oversized game of marbles. Close inshore, the brightly modelled faces of the Mackellars are just starting to show the first flush of evening, while to the east, the ice cliffs are already sliding into the deep ultramarine shadows of the new day. Total calm and absolute magnificence—to be anywhere else on earth would be second best.

As the evening advances, so the cold deepens but I still have the will to stay out. With the temperature now well below zero the surface of the snow, having melted during the day, freezes again and crunches satisfyingly underfoot. It has a texture all of its own, one to be savoured in the stillness. The sound of footfall on snow is not often heard in this place—from a squeak, to a rustle to a crunch it's usually obliterated by the wind.

Above the knoll near Gadget Hut, dozens of snow petrels wheel and dart, landing briefly by their nests, before flying off again in the same restless motion. The other creatures of the night, Wilson's storm petrels, tiny black apparitions, barely visible against the darkening rocks, flit about in silent nervous flight, hundreds of birds returning to the shelter of their nests, deep down between the boulders. By two o'clock I'm chilled and tired, and so reluctantly head for the tent. I had planned to stay out and see the sun rise over the shoulder of the plateau, but for now that'll have to wait in the hope that we're blessed with at least one more perfect night.

Sunday, 28 December

The day starts fine, but the katabatic which rose in the night blows throughout the morning at a steady 25 knots. A few among the group sleep an extra half hour or so, recovering from our Apple manoeuvres. But somehow I'm up, dressed and at the hut around nine o'clock, despite the exertions of yesterday and my twilight wanderings. Lying in bed listening to the tent flog is not my idea of rest. Anyway, the warmth and comfort of the hut is only a short, cold dash from the tent.

Rookie already has a large pot of water boiling on the stove. He's finished his breakfast, had a wash and is vigorously brushing his teeth when I burst in. Big, ginger-haired Rookie's very much a morning person and always first up. By the time any of the rest of us surface, he will have emptied the piss and kitchen slops drums into the bay, conducted a radio sked, done his weather observations and eaten breakfast. As cold as the air outside might be, Rookie's morning greeting is warm and enthusiastic. His badinage with most of us is often very silly and always a lot of fun. But mention the words 'sailing' or 'yacht', and be prepared for a half-hour talk about the boat he's currently building on the New South Wales north coast. Sailing is his passionate obsession. He's sold his house and land at Wooli near Coffs Harbour, and will some day soon just sail away in his beautiful fifteen-metre steel-hulled yacht. Destination: the Antarctic Peninsula, then every ocean beyond.

Rod and I collect water straight after breakfast. Someone's had the sense to sacrifice the water saucepan to the cause, bashing its rim into shape so it pours straight into the jerry can, without need of a funnel. While such improvements might seem trivial, they make a great difference when even the simplest task is a struggle.

Back at the Sørensen, Geoff and Estelle discuss the work program for the next few days. I join in, commenting that more photographs need to be taken of the interior of Mawson's Hut before the ice and snow are excavated. Perhaps set viewpoints for 'before and after' shots should be agreed on.

As the day advances, the wind abates. The afternoon settles into one of those rare, breathless calms. Estelle and Geoff take advantage of the moderating conditions and work outside Mawson's Hut, while the others press on with the Sørensen extensions. Meanwhile, I have the living section of the hut to myself.

Working alone inside Mawson's Hut is a heady experience. Despite the frigid surrounds, the space still resonates companionship and endeavour, as well as the ordinary desires of domesticity and comfort. Confined for long periods because of darkness, cold and appalling weather, the hut became more than just a sturdy shelter for up to eighteen men. Alone in the darkness and the quiet, I can almost hear the rowdy banter or a crackling gramophone, and see the cards flung on the table at the end of a game, or the smoke curling towards the rafters from a dozen favourite pipes. Mawson devotes a considerable space in *The Home of the Blizzard* to describing hut life through the course of two winters. Of returning to the hut, Mawson writes that 'passing through the vestibule and workshop one beheld a scene in utter contrast to the outer hell. Here were warm bunks, rest, food, light and companionship—for the time being—heaven!'

About 7.30 I return with Geoff and Estelle to the Sørensen, then we leave soon after for Gadget Hut and a meal with Don and Margie. In a gesture of great hospitality, our neighbours will be having all of our party to dinner in groups of two or three over the next few weeks. Joan and Rookie were their guests last night. For the first time in many days, the three of us enjoy sitting at a table to eat. Our dinner conversation turns to the McIntyre's plans. Don has more than one grand scheme he's working on. His latest preoccupation is a 40-metre yacht with an ice-strengthened hull. It'll be equipped with its own laboratories and a demountable float plane housed in its own stern hangar, and be available for charter to scientists and researchers. For Don and Margie, the new yacht represents a chance to spend at least part of each year exploring Antarctica. More than a couple of dreamers though, Don and Margie seem to be able to turn fantasies into reality by attracting significant support from sponsors. They may well succeed with this one, just as they did with Gadget Hut and their year together at Cape Denison.

We head for home about midnight. In our absence, work on the Sørensen has proceeded quickly. The wall panels have been stood in place and joined together, and the roof completed. By late evening it's done and universal relief fills the air. We arrive back to find the others enjoying a drink inside the new Sørensen. Dave Killick sports an impressive pad of white gauze over his right eye after being struck by 'shrapnel' working on the roof. Helping the other Dave rivet the roof panels together, a fragment of drill swarf flew into his eye at just the moment he wasn't wearing his glasses. The inevitable jokes about the one-eyed reporter follow. Even though he's a little subdued by his injury, the Killick humour still flows.

While the 'Sørensen Mark I' was a place of many a funny moment, it was also a source of growing tension as the cosy novelty faded. Joan suggests that Ted, Paul and Dave Gillott be awarded a Nobel Peace Prize between them for averting the first murder in Antarctica.

Monday, 29 December

The wind blows a hard 45 to 50 knots from south to south-east all day. Quite cold with overcast, sullen skies. Most of us have an easy time after the exertions of the last few days' construction efforts. Starting late, Geoff, Estelle, Dave Gillott and I together with the film crew head for Mawson's Hut to begin ice coring.

Geoff, Estelle and Dave are there to do the work; Mike, Mal and I go along to watch and record the action. Standing for hours at a time in the newly excavated workshop tunnel is cold and frustrating. There's barely room for the three workers let alone Mal with his camera and lights. Mike and I are relegated to just inside the outer door. Even though the others are just a few metres away, all we can hear are a few muffled sounds, mimicked by the flickering needles on Mike's Naga reel-to-reel tape recorder as it picks up the chatter in the tunnel via a remote microphone.

From time to time I shuffle past Mal and take a few quick shots of the ice corer in action. Made of fibreglass, it's simply a hand-cranked hollow auger. As the operator twists it deeper into the ice, further sections are added to its length. The cutting end is equipped with metal jaws. Dave drives the auger horizontally into the face of the ice accumulated in the workshop to a depth of about two metres. As it's withdrawn, a plastic sleeve is slipped over the core sample, ready for sealing and labelling by Estelle. It's hoped the ice cores will, when analysed back in Australia, provide information about snow accumulation and ablation within the hut and give clues about climate change since the workshop was last excavated in 1978.

With all feeling just about gone from my toes, it's time for a break. But I have more than a break in mind, so I head to the Granholm for 'bodily purposes'. Outside, a bitter, pitiless wind surges off the plateau and dark pewter-coloured skies droop low over Azimuth Hill. Large slabs of sea ice have broken away from Boat Harbour. The ice edge is now quite high and the fractured and deformed curve of the pressure ridge near our landing site is all but gone.

Although Rod and Joan, for medical and dietary reasons, do not entirely approve of the practice, several among the group seem to be able to exercise a degree of bowel control that allows them the luxury of forestalling the inevitable until they're in the vicinity of the Granholm. I'm fortunately one of them. It's actually something of a small pleasure to sit sheltered in the porch of the Granholm with the door ajar, dreamily looking out to sea or studying penguin society close at hand outside. In preference to the open-air dunny below the Sørensen, some have been known to make a special visit, a round-trip of more than a kilometre. Not very efficient I suppose, but infinitely preferable to the tail-chilling ordeal offered at the other facility.

Having not ventured further west than the Granholm, I decide to take more time out and climb the ridge to nearby Azimuth Hill while the others continue to fiddle with the ice cores. Steep approaches and absolutely no protection give this rocky spine an inhospitable edge. Small colonies of Adélies are scattered along the crest culminating in Azimuth Hill and the Memorial Cross. How they survive there astounds me. Near the top of the 25-metre-high hill the wind gusts at more than sixty knots. My approach is reduced to a cautious crawl for fear of being blown off the abrupt north-east side. I fight with zips, ventile hood, gloves and glasses to take a quick photograph before retreating. I'll return later to study the cross.

Back at the work site, Dave Gillott is a metre and a half down in the glacier just to windward of the Weatherhaven, making the final cuts with the chainsaw to a pit in which the ice cores are to be stored. His excavation extends into the blue ice, fashioning a rock-hard chamber in which the fridge box will lie buried until we leave Cape Denison. The compacted ice comes out of the trench in neatly sawn blocks. As I lift yet another heavy block, I ponder the obvious questions; how old is it and how far has it moved since Mawson's time? Cores drilled to great depths on the Antarctic plateau have been dated at close to 150,000 years.

We all lend a hand. The work is difficult in the torrent of wind, but Geoff and I manage to cart the excavated blocks one by one across to the Weatherhaven. We're all concerned for the future of this wriggling multi-coloured slug in the constant gales, so we pile the blocks on the windward side to further weigh down its valance. Arranged in a rough, low wall, the blocks will also help to deflect the wind around and over the Weatherhaven. With the cores safely inside, the fridge box is lowered into its glacial freezer and the hole backfilled with snow and ice. So the box can be found again, Mike places a marker stake on the corner of the excavation.

Tired and cold, we're all pleased to knock off and head for home. The walk back is hard, the wind ramming down the hut valley with a ceaseless monotony. I divert slightly to photograph the Weddell seals lying in a loose group near the ice edge. Like an old concrete path lifted by tree roots, the sea ice is riven with random cracks and pressure ridges. While the ice is still firm underfoot and not yet dangerous, it has a feeling of impermanence about it. In a matter of weeks it will be gone, consigned to the sea as part of the restless cycle of the place.

Joan's meals are always a treat. Her kitchen, whilst the haunt of thirteen restless individuals is also the scene of some remarkable culinary achievements. And it's all done on a domestic gas stove, there's no running water and the pantry is a jumble of black plastic bins outside in the snow. Tonight's meal is no exception. She serves a tasty vegetarian 'meat' loaf based on minced cashews which is wolfed down by all. The meal is well lubricated by more than an adequate quantity of wine and beer. Outside the wind continues to howl at more than fifty knots but for now, all's right with the world.

Tuesday, 30 December

Another cold and overcast day. After breakfast, procrastination sets in, so I look for indoor tasks. Cameras need cleaning and there are films to be logged. I'm unfamiliar with a fine old Hasselblad roll film camera lent to me by my trusting friend Hugo back in Australia, so I fiddle with its film holders for a while, trying to master their basic but not immediately apparent idiosyncrasies. In truth though, life in the outdoors doesn't look pleasant at all.

Joan reinforces my opinion when she returns to the hut looking distressed after a brief foray outside. Fortunately, Rod is on hand. He bundles her into a corner and gestures to everyone else to make themselves scarce so he can tend his patient. She had fallen heavily on the way to the open-air piss drum, skinning her knees and her dignity. Life is quite grim here at times. She's soon joking about it though, referring to the icy patch below the Sørensen on which she fell as 'the jaws of death'.

Eventually the appeal of inaction and confinement wears thin, my restlessness overcoming the inertia of a warm hut. With the newly mastered camera I head off towards the high ridge to the west of the Sørensen, intent on photographing the spectacular cloud formations in an everchanging Antarctic sky. The extreme wide angle of the Hasselblad's lens is ideal for the subject.

Everywhere between the Sørensen and Boat Harbour, aqua streaks are starting to appear in the permanent expanses of ice. With the slightly warmer air, frozen streams are slowly emerging from their hibernation, and as the surface changes, a milky, slick blueness warns of potentially hazardous stretches of ice. If the streams run at all, they will only ever form fleeting subterranean trickles. Where the winter cover is perennially thin—as in wind scours around dumps of glacier-rounded boulders, shattered rocks and flinty scree slopes—the snow and ice is rapidly receding. Like an emaciated foot, Cape Denison slips tentatively from under the hem of its robe of white. The bones of Antarctica are showing.

Cape Denison was of particular interest to Mawson. Being a rare ice-free expanse of rock, the cape provided him and Frank Stillwell, the AAE's geologist, with much information on the process of glaciation. Mawson and Stillwell made extensive rock collections and thoroughly mapped the cape. The predominant rock is a pre-Cambrian granite gneiss, in places greatly folded and deflected. Mawson found rocks among this parent strata that proved similar both in age and composition to outcrops in South Australia. Despite the evidence, Mawson, towards the end of his scientific career, was still not swayed by the emerging theories of continental drift and the great parent Gondwana to which Australia and Antarctica both once belonged.

With the power of the sky apparent in all directions, I'm content to go no further than the ridge above Gadget Hut. In the unobstructed view to the west past Cape Hunter, snow clouds gathering in huge baroque swirls hang low over the sea. Past the cape and on over the tapering western arm of the bay, where ice meets sky, the undersides of the clouds glow with an eerie pallid light.

Called 'ice blink' by the earliest sailors to venture into polar waters, this luminescence is caused by light reflecting off white surfaces onto the underside of clouds. Before any pack ice or the continent itself can be seen when approaching by ship, the horizon lightens, the crystalline surface of ice and snow scattering reflected light back into the sky. The phenomenon was observed by James Cook, the first explorer to probe south of the Antarctic Circle. At the end of January 1773, Cook noted:

> We perceived the clouds to the south to be an unusual snow white brightness which we knew announced our approach to field ice . . . [and] . . . the southern half of the horizon was illuminated by rays of light which were reflected from the ice to a considerable height.

Not content with just photographing the bleakness and tumult of the sky, I feel an urgency to press on with my paintings. I wouldn't like to return home with a sheaf of work executed just on fine days. That would give a very distorted view of the Cape Denison experience. It's time to test the level of my endurance. I need to be able to work for protracted lengths of time outdoors in whatever circumstances the elements dictate. Despite the wind, I feel that it must be possible to work anywhere, given shelter. After returning briefly to the hut, I head out again, this time with my painting gear.

A hundred metres to the north-east of the hut, past a boulder field of exquisitely rounded rocks, a small knoll seems to offer some prospect. A pair of South Polar or McCormick's skuas patrols the area, there being two small Adélie colonies nearby. Picking my way across the rocks I'm reminded of their presence by the sudden brush of a wing on my ear and a wild whinnying squeal.

Fiercely territorial, skuas are both the pre-eminent terrestrial predators and scavengers of Antarctic shores, the scourge of Adélie penguin colonies. During the breeding season their livelihood derives overwhelmingly from Adélies. Arriving for the summer, before the penguins, they initially survive on last year's 'left overs', or, as Don McIntyre described them to me, 'frozen dinners' of dead adults and chicks, and juveniles that didn't survive the moult. Once the Adélies have mated and eggs are layed, the skuas are always ready to pounce on unprotected eggs or chicks and will harry and kill sick or injured adults. Other birds, particularly snow petrels, are taken and skuas are quick to scavenge on seal carcasses and any carrion on offer.

My prospective place of shelter turns out to be a penguin-moulting station. The rocks all about are smeared in various shades of cream and pink guano and dusted with a crop of fine, moulted feathers. A distinctive smell, vaguely like ammonia mixed with stale fish, is heady enough to persist against the breeze. Not the most attractive place to work, but such is the lot of the itinerant Antarctic artist.

The fast ice around this stretch of coast is retreating quickly. On a small remaining strip, a white shape moves behind a lump of ice. The skua that had previously swooped at me glided to the ice, harassing the intruder, a southern giant petrel. The only other terrestrial predator in Antarctica besides skuas, the petrels are predominantly offal and carrion feeders but will also attack and kill other birds when in residence at their breeding grounds. While southern giant petrels inhabit the shores of Antarctica, they are very much the garbage collectors of the sub-Antarctic islands.

I paint on until light snow flurries render it an increasingly futile activity and return to the hut, cold and stiff in the fingers. In my absence a disaster is averted over at Boat Harbour. What we'd thought to be the highest crack in the ice foot is succeeded by another that comes to within a metre of a large pile of timber. Once firm ice that we'd walked on just in the last few days has now broken away and drifted out of the harbour. A hurried detail of four or five workers moves the timber to a point almost parallel with Mawson's Hut.

The wind's up again in the late afternoon, blowing about sixty to seventy knots from the east. There must be a low-pressure system close by for us to be experiencing winds from that direction. With everyone back inside by nine o'clock, who cares which direction the wind's coming from? The old and the new Sørensen are now joined! During the course of the afternoon, Dave Gillott and Paul, assisted by Rookie and Alan, cut an opening through the southern wall of the original hut and our lives are thus transformed. Dominating our 'dining room' is a new table, large enough to seat thirteen at a squeeze. Its top a sheet of plywood forming, this magnificent piece of furniture is just a small something Paul knocked up in a couple of hours.

With all these domestic improvements the mood over dinner is celebratory. The day, that for Joan started grimly now ends positively; no longer does she have to share her kitchen with twelve others. Also, Joan's ordeal of this morning prompts a communal decision to give over the Apple entirely to 'civilised' toiletting and bathing. Perhaps we males have been a bit slow in fully appreciating the awfulness of the open-air arrangements where Joan and Estelle are concerned.

As we eat, Geoff gives us a summary of the day at Mawson's Hut. With only Ted and Estelle assisting, he's a little disappointed in progress but reports that the ice cores have been taken from above the ceiling platform in the living section. Full-scale excavation can now begin in the living section with only the final ice cores through the roof hatch of the workshop yet to be drilled. I detect that Geoff's a bit resentful of the attention the 'home camp' is getting. He's keen for Dave Gillott and Paul to finish at the Sørensen as soon as possible and return to Mawson's Hut.

I enjoy a few beers until about 1.30 then head off to the Apple and then the tent. The stayers—Paul, Ted and Alan—look set for a few more hours.

Wednesday, 31 December

The day begins well as we're all now able to enjoy a more leisurely and spacious breakfast without sitting in someone's lap or getting under Joan's feet. By late morning the wind eases from 60 knots back to 15. Rookie records a nine o'clock temperature of −2°C.

I walk over to Mawson's Hut shortly after Geoff and Ted to assist with the ice cores and general recording. Distractions abound along the way. The ten- to fifteen-minute walk is, I find, often taking up to half an hour. Ice continues to slip away from the foot around Boat Harbour. As a consequence, the Weddell pack has retreated further and are now often close to our path across the flats. Their hides vary greatly: from a silky light-grey of the calves to beautifully mottled coats ranging through many shades. Moving among them hardly draws a response. A lazy lift of the head, heavy sigh or the opening of their large liquid-brown eyes, and they quietly return to their slumbers.

At the hut more ice coring is in progress, this time as planned through the workshop roof hatch. The job of manipulating the ice auger while perched on the roof falls to Ted and Geoff, hand-cranking the beast through the recently removed skylight. Once the core has been taken and handed to Estelle for bagging and labelling, the chainsaw is again fired up with Ted returning to the work face inside the hut. With the chainsaw, Ted makes the tedious process look easy as he pushes a branch tunnel into the main body of the workshop.

Geoff is keen to examine the structure of the workshop roof from inside. Just as has occurred in the living section of the hut, the ceiling platform in the workshop has collapsed, fracturing the collar ties in the process. If functioning properly, collar ties should stop the walls and roof from spreading, but it's likely that in their current state they're structurally ineffectual. Geoff and Ted probe upwards through soft snow and find the ceiling space almost full to the brim. Both are not enthused by the prospect of excavating all that snow just to effect repairs to the collar ties. A rethink is required.

The problem is discussed over lunch, but Geoff seems to have already made up his mind. He concludes that the planned recladding of the workshop should brace and stiffen the roof to some extent, and the mass of snow and ice within the workshop is likely to act as a solid tie thus preventing the building from spreading or twisting. Finally, it's decided that under the circumstances, it would be better to direct our energies to excavating the living section and leave the interior of the workshop to a future work party.

Mal and Mike are also at the hut. Together, the three of us labour for a while assisting Geoff and Ted by removing blocks of ice from the workshop end of the tunnel. The accumulated blocks are then dumped on the growing windbreak behind the Weatherhaven. We're all relieved that the Weatherhaven is still intact. Apart from the storage and shelter it affords so close to Mawson's Hut, the thing is becoming something of a social haven, particularly when lunch arrives. Yesterday, Ted patched some small holes where its fabric had chaffed through and restrung a number of guy ropes, anchoring them to very effective deadmen in the form of full gas bottles.

Eager to take advantage of the increasing calm, I return to the Sørensen, then walk down to our little bay for a painting session before dinner. Despite the light breeze, the air is very cold and I have to rug up, leaving only my nose exposed. The strand line on the ice is littered with piles of kelp. Motionless among this rusty fringe lies a southern elephant seal, probably a young male. Though not common along these coasts, such a sighting is not unexpected. Elephant seals are predominantly creatures of the sub-Antarctic, where, at the end of each winter, thousands haul out on island beaches to fight, mate and moult. Mature males engage in bloody combat with one another, their prize being dominance of a stretch of beach and its resident harems of cows. Unsuccessful or immature males that haven't made the grade as 'beachmasters' are consigned to frustration. In roaming the Southern Ocean they often haul out in summer on the Antarctic ice, either singularly or in small groups. No elephant seal breeding occurs in Antarctica.

Pleased with my efforts, I return to the Sørensen after about three hours to find Paul and Dave Gillott continuing to transform our Antarctic house into a home. Along the length of the eastern side of the hut extension Paul has built a work bench, topped by two banks of shelves. In the corner closest the door we now have a breakfast bar and small pantry, courtesy of Dave. It's already stacked with cutlery, plates, a row of mugs on hooks and tomorrow morning's breakfast offerings. Dave has even made a portable toilet seat for the Apple.

The work bench and shelves will be a boon to productive work for those who need the space. And now I have a studio. It comprises only about a metre of bench space that I can legitimately call my own, but there's a view directly above, and it's one that I'd cherish from any studio. As I organise my camera gear and art materials in the studio, I pause to look out the window at the distant scene, a small part of which I had painted not an hour before. In the middle distance a squat dark shape lumbers down off the band of dried kelp, across the ice foot and into the bay. The elephant seal has at last moved.

Another vegetarian meal tonight. Albeit very tasty, some people are starting to mutter about the perceived imbalance in our diet. After dinner, Dave Killick reads the latest news received from AAP via his computer. Of course with Dave reading, the most tragic reports of human misery or misfortune are rendered humorous. Before long we're nearly all crying into our drinks with laughter. He is so warped, but wickedly funny. The McIntyres arrive just before midnight and together we go through the annual ritual. But the New Year's celebrations are subdued—accumulated tiredness and the rush to finish our home away from home have taken their toll. We all drift away early to savour our own thoughts on the new beginning.

Thursday, 1 January

The night passes in peace. Outside the tent there's just the brush of a light breeze to accompany the welcome return of the sounds of life all about. Someone next door snores. As the sun returns to Cape Denison a rare and gentle day appears to be in the offing—such a propitious start to a new year.

About mid-morning, I head off in the company of Mike and Mal for the cliffs of John o'Groats at the eastern extremity of the cape. I have the strong feeling that after so much trying weather this day will be made for venturing further from our camp than any of our party has to date. We step out into a perfect Antarctic morning, with a barely perceptible breeze drifting off the ice to add a pleasant crispness to the air.

Whether from nostalgia for the mother country or simply through a whimsical sense of humour, the opposite ends of the two-kilometre-wide peninsula were named by members of the AAE after the extremities of the British Isles. I've not been to Scotland, let alone the original John o'Groats, but I'm sure it's nowhere as much the point of no return that each of Cape Denison's rocky bounds are.

The entire journey is filmed, so a walk of barely a kilometre takes the best part of two hours. Both Mike and Mal seem relieved to get away from the toil and serious endeavour associated with work on Mawson's Hut. In this excursion they see the potential for filming abundant wildlife, spectacular scenery and a member of the expedition not fighting the elements or engaged in heavy labour and building.

We do several takes of me coming and going, scanning the horizon, walking up, down and around the rocky ribs of Cape Denison. I'm depicted as the artist surveying his domain, searching for a subject. Portfolio under one arm and sporting a backpack, I climb the hills, pausing from time to time, broodingly scanning the horizon in search of 'prey'. My feet attract particular attention. Mal shoots take after take with his compact digital video camera attached to a boom. He skilfully tracks my footsteps, with the camera passing centimetres from the ground, as I crunch across loose scree slopes or amble close to one of several Adélie colonies on the way to John o'Groats.

The extensive screes on the descent to Low Lake are an unexpected delight to the three of us. They're not just unstable jumbles of sharp rock fragments, but more like a geological garbage dump filled with treasures for the eager eye. What makes them remarkable is the endless variety of gorgeous individual stones—reds, greens, sparkling blacks and greys, some banded and striped, from tiny flints to car-sized boulders. Known as erratics, they are characteristic of the terminal moraines found at the foot of glaciers or ice sheets. All have been stripped from their parent rocks in unknown places and depths, transported perhaps hundreds of kilometres and eventually disgorged by the ever-moving ice. Mawson described the terminal moraines of Cape Denison in *The Home of the Blizzard* as 'happy hunting grounds'. Here, he said, 'the story of the buried land to the south is in large measure revealed in the samples brought by the ice and so conveniently dumped'.

The valley before us catches a receding tongue of the ice cap. In places its floor looks slumped, indicating the possibility of crevassing, so we cut towards the plateau, then cross close to the top end of the frozen lake. On the eastern side lies the outer suburbs of a large Adélie colony that appears to stretch all the way to the ice cliffs. Perhaps twenty thousand pairs frequent this rocky stretch of Cape Denison.

The Adélie penguin was named, as was the coastline it inhabits, for Adélie, wife of Dumont d'Urville. Adélies are the most numerous of the Antarctic penguins with nearly two and a half million pairs estimated as breeding around the Antarctic coastline and its immediate islands. From March to October they feed off the edge of the pack ice and are gone for the duration from their nest sites, only returning to shore in the summer months to breed and moult. They don't have much time, for by the end of February the Cape Denison colonies, like those all around Antarctica, will again be deserted.

At present, the breeding cycle is in full swing with most adults tending chicks, many of them two. The chicks would have been conceived in October and hatched in November or early December. At their current stage of development, the chicks remain under the constant care of one parent while the other is at sea feeding. In the ensuing weeks of January, as their down is replaced by feathers, the young birds will gather in protective crèches and by seven or eight weeks will be independent of their parents.

They look very fragile and probably are. Of the thousands in the John o'Groats colonies arrayed before us, large numbers will never go to sea, the victims of skuas, blizzards, starvation and neglect. But there's strength there also. Even at a distance, a thin chorus of tremulous 'cheep cheeps' from thousands of chicks can be heard beneath the raucous braying and calling of the adults. Singularly it may be a tiny, weak voice, but collectively it is the persistent cry of survival.

On finally reaching a suitable spot for painting, I'm overawed by what surrounds us. The ice cap looms large and blue behind, its slowly arching skyline brushed by wisps of cloud. Out to sea the familiar markers, the large saddle-backed berg and the Mackellar Islets, all ride easily on a languorous calm. Directly below, steep terraces of penguins descend almost into the sea and all about us, penguins in their thousands go about their business in near total indifference to our presence, only emitting a guttural growl when one of us strays too close.

Over the brow of the final knoll, our tenuous foothold in this alien place ends suddenly as the ice cliffs begin. Every shade of blue and grey facets and moulds the shadows of their forbidding heights. Unapproachable yet alluring, they look dangerous, their enormous jutting cornices ready to drop at any time. A few hundred metres across the nearest bay from our vantage point, showers of ice fall from the roof of a large cavern as deep rumblings reverberate from within. Seconds later, a bluish-white vapour is exhaled from its mouth as a tide of loose brash ice starts to drift slowly seawards from the cliffs.

I'm both relieved and enlivened to be able to work in such calm and almost balmy conditions. Mike feeds me chocolate, then begins to talk as I set palette, brushes, water and paints around me, placing my equipment wherever the rock will allow. With only a gentle breeze, there's no need to be on constant watch against paper, palette or portfolio being blown into Commonwealth Bay. As I start work, Mike asks me about Antarctica, Mawson and my reasons for being here. Mal's camera rolls. My answer in part is a wide smile and a simple expansive gesture to all that lies about us. How can it be encapsulated in just a few sentences without recourse to the usual superlatives? Antarctica is a word-robbing experience. All I can say is that I hope my paintings might eventually complete the answer to Mike's question.

The painting finished, we explore the lower terraces of the Adélie colony, gingerly threading a path past thousands of penguins. Fortunately, we find a steep gully free of nesting birds over the most difficult stretch of the descent. It's messy going though, as the slope is a running sewer of rich and putrid effluent, drifts of feathers glued to the rocks, and carcass after desiccated carcass of those that didn't make it. At the rock edge a skirt of snow and ice marks the end of our travels. From there any slip would mean death within minutes. The water, like the great expanses of ice, looks enticing yet is utterly out of reach.

Curtains of deep blue begin to close over the ice caverns and unattainable grottos below, but up on the terraces the afternoon sun still graces the assembled throng. On a steep north-facing slope leading to the edge of the glacier, a congregation of perhaps two thousand Adélies stands scribed against the rock by the crystalline light, their white breasts shimmering like pearls. Many are tending chicks. Wedged between feet and belly, some small chicks barely protrude from under a blanket of parental brood feathers, while larger chicks wriggle and squirm at the adult's feet, constantly craving warmth. Where there are two, the desperate need for body heat is, next to food, the dominant theme in the struggle between siblings, the stronger of the pair enhancing its chance of survival by pushing closer to its parent. Downy grey, all flippers and feet, the chicks at this stage of their growth aren't able to stand and so shamble about like tiny inebriates.

On the return journey to the Sørensen, Mal and Mike go on ahead while I pause to paint again, this time taking in the glories of evening light. The painting is just rocks and the little bay next to the east of the one below our camp, but like the morning's work it flows freely and easily. Sprawled on a north-facing slope of large tumbled boulders, the late sun reflecting off steel-grey rock, I feel warm, drowsy and at peace. A fitting end to a rare and special day.

Back at camp, Joan has prepared a New Year's feast of great proportion—two sides of pork complete with crackling and all the trimmings, more than enough for our group, plus the McIntyres. That should redress the balance in favour of the carnivores for a while. There's liberal quantities of drink too, including a six-litre bottle of red from our sponsor, Yalumba. Once again Dave Killick is in good form, with his jokes flowing as freely as the wine. We're all at a loss to know how and why he devotes brain space to such an extensive repertoire of tasteless jokes; and where did he hear them? Despite his relative youth, this one-time hotel chef, crime reporter and peace protester has knocked around. As a crime writer, Dave's seen more than his share of the seamy. Among the stories he regales us with those of being sent by pushy editors to harass murder victims' relatives for photographs of the deceased confirm all our worst prejudices of the tabloid press.

Via satellite Dave has been sending back a steady flow of stories and digital images to AAP in Sydney for distribution through the Australian media—a difficult task that's tested both his patience and ours, but in particular Rookie's. Now messages are coming back demanding more. The pictures that we all thought showed the drama of Antarctic life to great effect are not always pleasing Dave's superiors nor our sponsors.

One request though is a source of great mirth. The arms of our ventiles are emblazoned with sponsors' badges that are obviously intended to be caught in Dave's or my photographs. Headquarters are complaining that the sponsors' logos are not featuring clearly or prominently enough in the digital pictures and one side seems to be favoured over the other. Back in Sydney, a commercial photographer is consulted. He recommends that group shots of daily life and work be posed. Worst of all, he thinks everybody looks too serious! So the message comes through, 'can we remember to keep our arms straight and have both showing in the picture. And don't forget to SMILE!!' If only they knew how hard it is to work in the cold and wind, let alone compose a happy, smiling photograph. Someone suggests that we pose the obvious shot of a soldier-like row of straight-armed figures—smiling of course—and see how that's received back in Sydney.

The celebrations and hilarity continue well into the morning.

Friday, 2 January

A morning meeting is held to review our progress so far. Predictably, it runs on into the early afternoon. While a great deal has been achieved—the Sørensen extension, relocation of the Apple and work inside Mawson's Hut—Geoff for the first time gives full voice to his concerns about progress on the conservation program. He says that we can't afford any delays caused by splitting resources and continuing to work at the Sørensen camp as well as Mawson's Hut. Others say they want a regular day off and consistent working hours. Joan gives this idea her weighty endorsement; partly so she can plan meals efficiently, but also in recognition of all the tired faces seated around the table. It's proposed that work start for the day at 9 o'clock and that dinner be at 7.30 each evening. I know I'll sound like a wowser, but with the late night and early morning drinking sessions I'm not confident that this regimen will be kept. It's all very well to talk of regular time off—let's get some discipline into our days first, then it might be achievable. Anyway, it's probably better to think about rest when the project is better advanced. The meeting also agrees that a formal domestic roster should be drawn up covering washing up, water collection, and kitchen slops and piss drum emptying.

I freely admit that I didn't voice an opinion during the meeting. I'm in a unique position here, with a work program that's entirely my own. I can plan my time according to my objectives and not the group's, and in this circumstance I see great advantages but also responsibilities. My work succeeds or fails on my efforts alone. Also, apart from the labouring I might do and the cold and wind I endure, my tasks are a pleasure, a total delight. So I feel that it's not my place to suggest to the others how they should function while in Antarctica. However, I'll continue to consider routine days off and even regular hours a bit 'work to rules' for this place. We're only here for a short time. Within the limits of our own stamina and safety, I hope we wring every last drop out of the experience.

After lunch, Alan and I walk over to Mawson's Hut together. We enjoy each other's company and I think he's becoming a good friend. I like his style of leadership. It's not prescriptive or formal. Alan sees himself as a coordinator and organiser rather than someone leading from the front. He'd like to be thought of as simply one of the blokes, who, for his sins, bears some added responsibility.

The day is overcast and the wind only blowing about thirty knots, so it's hoped that work on the outside of the hut might finally get under way. We arrive to find Geoff and Paul already on the hut roof attempting to remove scraps of ridge capping from the workshop. Once this is done, it's envisaged we'll be able to carefully strip the existing badly weathered pine boards from the entire workshop roof. New boards of a matching profile will then be installed before the old are reinstated over the top, in as close to original fit as possible. Quite a task. No one has any illusions as to its difficulty.

Necessity being the mother of all Antarctic solutions, the ridge caps were actually fashioned from old biscuit and fuel tins, the intended metal sheet missing among the stores landed at Boat Harbour from the *Aurora*. After more than 80 years of exposure, the capping has worn thin and is very brittle, not so much through corrosion but ice abrasion. I climb onto the roof where Geoff and Paul are in earnest discussion over these fragile pieces of iron. Their faces show that all's not well. The capping sheets are held down with numerous flat-headed nails. They prove almost impossible to draw from the dry timber without their heads breaking off or the iron being further damaged.

Paul tries drawing nails from the roof cladding itself. The same problem results, as this time the desiccated timber splits. It's obvious that our plan for the workshop roof is quite impractical. Even if it were possible to remove the old cladding without doing significant damage, the task would be slow and tedious, attended by the certain havoc of blizzards and high winds. There's nowhere near enough time—the whole summer might be needed for such a job.

In the living section of the hut, Ted has made significant incursions into the massive load of snow and ice burdening the ceiling platform. It's cold and wet work. As it cuts, the chainsaw spurts a slurry of melting ice all over Ted's chest and legs. Ventiles are fine for the winds outside but in the warmer air of the hut, he's getting wet, cold, stiff and tired. Whether because of his undoubted skill or because no one else is fool enough to try, 'Chainsaw' Ted seems doomed to be the one to complete this onerous job. Alan and I help ferry blocks down the tunnel and I photograph the work in progress before returning to the Sørensen at about 7.30.

Walking across the Boat Harbour sea-ice flats, Geoff, Estelle and I notice a stranger in town. The elephant seal has moved from the bay near the Sørensen and is now hauled up on the ice among the scattering of Weddells. For their part, the Weddells seem totally unmoved by the newcomer's intrusion. With easy access to the water, it's obviously a favoured haunt of all comers. Mawson records an occasional crabeater seal here and in *The Home of the Blizzard* recounts the presence of a large male 'sea-elephant' in the vicinity of Boat Harbour in February 1912. Measuring 'seventeen feet six inches in length and twelve in maximum circumference', the giant was discovered challenging one of the dogs. Duly shot before it could return to the sea, it was then butchered. The resulting 'mountain of flesh' was added to the stockpile of dog food, while the skin and skull went to the biological collection.

Living 'off the land' was not only an accepted practice but considered vital to the survival of man and dog in the Antarctica of Mawson's era. However, the killing was not profligate. The resident wildlife was just an opportunistic resource, there for the taking, its slaughter a grim part of Antarctic life.

Seal was regularly on the menu at Cape Denison, appearing in various guises from stews to sausages. Seal blubber was used as cooking fat and the offal eaten too. Penguins and penguin eggs were also consumed. These fresh supplies supplemented the otherwise limited and monotonous diet of tinned or dried foods. The dogs survived almost exclusively on seal meat and blubber. As the availability of seals and penguins was seasonal, considerable stocks of both were laid up before each winter.

Everyone is back at the Sørensen by eight o'clock for our first meal at a 'normal' time. Needless to say there is no penguin on the menu, nor will there ever be. Eating the wildlife went out of fashion years ago. Joan serves the usual marvellous dinner. Tonight it's a North African spicy vegetable dish served with cous cous. Unfortunately, it leaves me with an extreme bout of indigestion. It's not the first time I've suffered on this expedition and from the anguished faces and mutterings of others we might be in the midst of an epidemic. Dave Killick, Ted and Paul have complained of it. For Ted it's a new and unwelcome sensation. Joan is quick to blame chocolate, alcohol and the 'dead things' as she unfortunately insists on calling meat, but I think it's her beloved bread! I don't feel our domestic gas cooker is quite up to the task and I suspect there's some uncooked yeast being eaten. But as it's so delicious, I'll still eat the bread and suffer the consequences. For that matter I don't think I could cut down on Joan's suspects either.

Between about eleven and midnight the evening sky is at its best. Over the eastern end of the bay a hole has persisted in the clouds for the past few days: a rent of blue in draperies of grey. It's still there, now pale aqua wreathed in grand flourishes of a warm, dark grey with lighter pinks and creams embellishing the horizon. Inevitably, I'm drawn down the hill with paints and paper. I make a good start. The air is quite still, but it's as cold as I've yet endured while working. Soon the paint becomes gritty as it begins to freeze and my water pot starts growing icicles. The addition of some antifreeze to the water seems to help and I paint on for a while, quite pleased with my response to the sky. Finally, after about an hour and a half, cold and my indigestion win out with the work half finished. I head straight for the tent, my sleeping bag and the rather unpleasant chalky tablets that Rod gave me to chew.

Saturday, 3 January

The morning starts with an urge that must be satisfied. It's now a week and a half since my last full wash, and as the weather is reasonably clement, today has to be bath day. What convenience! No longer does one's ablutions involve a cross-country hike with a round trip of nearly two kilometres. Our bath house is just next door. After the big boil-up on the Sørensen's stove, I luxuriate with a dish of hot water in the sanctuary of the Apple. In its dome-like confines, alone and out of the wind, enjoying a bath almost assumes a hint of the spiritual. Not so my washing but even here I enjoy a small sense of achievement from the sodden little pile that I take from the Apple. Wearing clean underwear and socks, I now feel quite virtuous and in a strictly jocular way I can be superior to the others, if only for a short time.

After spreading my long johns, underpants and an assortment of wet socks on the rocks, well weighted with small boulders, I organise cameras, film and painting gear for the afternoon. A scratch lunch of biscuits and cheese hurriedly eaten, I head off towards Boat Harbour. Rather than going first to Mawson's Hut, I decide to investigate Penguin Knob on the eastern arm of Boat Harbour.

The flats extending out into the bay are littered with AAE remains, much of it blown from the other side of the harbour. Persistent fragments of timber, old tins and wire lie wedged between the rocks. As we tend to concentrate our efforts around the main hut and the scientific observatories, it's easy to forget that the whole of Cape Denison provides evidence of the AAE's activities. Mawson and his men combed the whole peninsula and surrounding waters gathering geological samples and biological specimens, making photographs and magnetic observations, mapping, surveying and gathering food.

At the northern extremity of Cape Denison, Penguin Knob offers a fine view back over Boat Harbour to the hut valley and the ice plateau behind. From such an angle the ever-rising ice looks like a great wave, slowly gathering energy before unleashing its tide. It seems portentous, threatening, ready at any moment to sweep away all traces of human endeavour from this place.

The knob itself is dominated by wonderfully eroded mounds, bulbous nodules and warty lumps, sliced through from north to south by several dykes—known to geologists as mafic intrusions. A geology lesson of fire and ice, these distinctive basalt dykes, as straight as tram tracks, have been squeezed as molten rock into the host granite gneiss up to 2,000 million years ago. Long since, ice has worked slowly on both formations, scouring and clawing at the gneiss and shattering the exposed surface of the dark grey dykes into incalculable knife-edged shards.

There's little protection out on the knob from the 30-knot southerly. Painting shelter is hard to come by, and as mentioned before, is inevitably at the cost of sitting in penguin filth or in the company of a weathered carcass or two. I finally make the best of the location by settling on a well-used benched terrace of guano-smeared rock only about five metres above the water's edge. My subject looks to the west over the long spit tapering from Azimuth Hill to the ice cliffs beyond. The Memorial Cross is prominent on the skyline.

It's a gloomy afternoon, the biting wind scudding across Boat Harbour and straight in my face. I keep lurching this way and that, catching the lifting edge of my portfolio, drawing board or palette before they're whisked out of my grasp. I'm distracted and cold and can't make a confident start. The composition of the picture doesn't work and the more I try to make corrections, the more problems emerge. Continuing seems futile, so I pack up rather dejectedly after just an hour's work.

While painting, I'd been diverted by the sight of four figures climbing over the rocks from the Granholm towards Azimuth Hill. Work was obviously about to commence on repairs to the Memorial Cross; reason enough to abandon my painting and record the activity on the hill. Leaving rucksack and portfolio at the Weatherhaven, I then head to the steep hillside and clamber over the boulders to the cross, cameras in hand.

As the highest point on the ridge running from Land's End to the northern extremity of Boat Harbour, Azimuth Hill commands one of the best views on the Cape Denison peninsula. Directly to the west the ridge drops sharply to the open waters of Commonwealth Bay, while to the south beyond the bare rock, the full curve of the ice plateau scribes the horizon in a sweep through 180 degrees. The theatricality and grandeur of the site must have made it an obvious location when Mawson decided to build a memorial to Ninnis and Mertz. All that remains standing of the cross erected in November 1913 is its upright, the horizontal arm having blown off in the intervening years. Even as a lone stick, it's prominent on the skyline and is one of the first man-made features visible when approaching the cape by ship. As part of our work the cross arm will be reattached.

I arrive to find that Paul, Geoff, Rod and Dave Killick have managed to free the dislodged arm from ice at the base of the hill's final step, just to the north of the cross. Paul has also begun fixing specially fabricated stainless steel brackets to the upright, preparatory to reattaching the arm. The scene with a ladder, three figures and turbulent grey sky beyond looks like an old master's Antarctic variation on the descent from the Cross.

Just below the brow of the hill, snow petrels and Wilson's storm petrels dart about. Rod locates a pair of 'snowies' which I have great delight in photographing. Their nests are not obvious, always secreted away in deep clefts among boulders, out of the wind and away from the eye of the skua. The first indication of a pair nearby is a muffled subterranean cackling.

Back at Mawson's Hut, Ted has installed a chute made of form ply that runs from the western skylight opening in the living section, across the roof and over the adjacent artefact field to the snow bank outside. Simple but brilliant. Rather than man-handle loads back through the tunnel inside the hut, gravity can do most of the work. Now, as blocks of ice are cut from the platform they can be shot directly down the slide, then tossed onto the ever-growing pile behind the Weatherhaven.

The day's work ends about seven o'clock with Alan and I walking back to the Sørensen together. The fifteen-minute trip to or from Mawson's Hut always affords a chance to reflect on our experiences, to chat or dream, or to simply observe the restlessness of Antarctica. Always different, today's crossing is no exception as Alan and I have a rather too-close encounter with the elephant seal. We stop for a photograph, also attracted by a very comely Weddell seal nearby. Alan lies down on the ice for a low shot. In an instant, the seal is out of the wallow that its body heat has melted in the ice, and is heading straight for my camera bag and portfolio. It stops and we relax, resuming our photo session. Suddenly it's off again, coming to rest within a metre of my camera bag. Alan manages to grab the portfolio, but what of the thousands of dollars worth of cameras in imminent danger of being flattened. I dare not approach. Then it moves again, its great flabby chin and slobbering mouth coming to rest on the bag. 'What do I do!' I exclaim to Alan, half laughing and half in a plaintive call for help. Alan waves the portfolio, creating a diversion. The animal instinctively veers off in his direction. I quickly make a grab for the bag and we both flee laughing, but more than a little relieved that one of us was not bitten, or my precious cameras steamrolled. Back at the Sørensen we receive little sympathy from Joan. She claims that we asked for it—molesting what she asserts was a female.

Another still, overcast night. Since a day or so before New Year, the wind has been nowhere near as blustery as in the first week. Going to bed has become a pleasure as I can, for a while at least, enjoy easy slumbers without a constantly flogging tent. Penguins bark and call outside and I can actually hear the sea.

Sunday, 4 January

An easy start today for most as this is their day of rest. Not so for me. There's only a light breeze blowing, so work beckons.

After a quick lunch, I head for the Memorial Cross as I'm keen to describe the view first seen yesterday as I looked west to the ice cliffs immediately past Land's End. Just below a small saddle running north from the cross, I find shelter on a broad west-facing slab. Here I'm protected from the southerly breeze by the boulders crowning the hill itself.

Close by in their stony grottoes, 'snowies' cackle to themselves. The western slopes of Azimuth Hill are home to a loose colony of snow petrels, one of a number scattered across the boulder fields of Cape Denison. The total population numbers perhaps twenty pairs. Steeper slopes are preferred as nest sites as there's less chance of their entrances drifting up with snow.

If the pair nearby has an egg, it won't be far off hatching. One egg is laid in late November or early December, hatching about forty days later. Snowies are to me the quintessential Antarctic bird, their name couldn't be more apt. In the nest, the vivid, almost fragile simplicity of their white plumage and sparkling black eyes and bill belies the robustness essential to any animal attuned to the rigours of Antarctica. Despite particularly heavy predation from skuas, snow petrels are splendid survivors.

The air below my perch is busy with birds. Snowies wheel back and forth in repetitive fluttering patterns, occasionally landing near their nests while hundreds of Wilson's storm petrels flit and glide in easy passes. Always busy the 'Willies' never seem to land.

Beginning at the bottom of the slope, and separating my painting and me from Commonwealth Bay and the ice cliffs beyond, is an expansive network of penguin colonies. Rather than nesting in one large aggregation, the Adélies nearby gather in groups of up to several hundred. On the next ridge across before the bay, these sub-colonies scatter over the landscape like little hill towns. Each makes its own mark on the hillside, the guano of so many birds over countless generations forming a dull pink pavement over the rocks.

Running up the gully and connecting colonies to left and right is a snow-lined main thoroughfare, the northern end of which spills onto gently shelving rocks just above the tide. There's constant activity on the street as loosely organised lines of penguins shuffle back and forth through the dirty snow. Some birds invoke the help of gravity by flopping on their bellies and tobogganing on the downhill runs, only churning with their feet when overtaken by their ambulatory companions.

Pleased with my three hours' work, I pack up, then scramble down the slope to investigate the gully and penguin highway close at hand. No sooner have I turned south up the snow ramp than the whinnying warning call of a skua pierces the air close by. There's a sudden whoosh from behind. I duck instinctively, half twisting my head to see its rapier-like bill just overhead. A few more steps and another bird swoops. They seem to be a particularly aggressive pair as I'm not near any rocks and so couldn't be an immediate threat to their chick. But their bullying tactics are very effective. Backtracking, I walk around a small point to the west, then along the ice foot and a narrow ramp of snow leading up from the water's edge. Here, I settle about fifteen metres above the shore and work on my second painting for the day, this time looking back along the increasingly precipitous cliffline that terminates in Land's End. In a way, I win out over the skuas. After commencing work, I find that this snowy road is also used by the penguins. The traffic is not as heavy as in the gully, but I'm rewarded with the occasional, if not very cooperative, subject for the foreground of my picture. There's a curious one in every crowd, and an occasional passer-by diverts from their path to inspect me and my gear. Over at Mawson's Hut, the Adélies that often stop to look at what we're doing have been dubbed the 'building inspectors'. Perhaps I've encountered the 'art critics'. At least they're not interested in lingering or looking over my shoulder.

I return to the Sørensen around seven o'clock well pleased with a very productive day. It's Joan's night off according to the new regime, so Ted organises a very acceptable freeze-dried meal of beef and something with rice. As a consequence of the general day of rest, there are a few people sporting 'fluffy hair'. It seems the Apple was a popular spot.

Monday, 5 January

Rookie's situation report records a 30-knot wind for most of the day and a maximum temperature of 1.5° Celsius.

Repairs to the Memorial Cross continue to prove troublesome. The prefabricated brackets don't quite work in situ, so Geoff and Paul fabricate additional washers and packing shims from galvanised and stainless steel plate. Also, working in an area remote from power tools means that materials must either be measured and fitted up at the cross then taken down to the Weatherhaven for cutting, or power has to be provided on top of Azimuth Hill. The latter seems preferable, so in relays we hump the small generator over the rocks to the cross. Then of course, we don't have the correct drill bit, or some such item, requiring another round trip of 20 minutes. And so the afternoon drifts by.

Rod discovers a small tin among the rocks just below the cross and Estelle's on hand to assess the find. Badly rusted but still with its lid intact, the oblong tin had an embossed pattern stamped on its underside, so was obviously a matchbox complete with its own striker. Despite my suggestions that it might best be removed and returned to Australia for conservation, Estelle is adamant that good archaeological practice be followed and that it remain where it was found. I duly do the photographic honours while she makes some notes, before returning it to its resting place. I suppose out here it's much easier to imagine it tumbling unnoticed from a pocket of one of the many AAE pipe smokers, some blustery day in 1912. Context is everything.

Some of my most enduring memories of Antarctica are of the myriad colours of the sea: from burnished sheets of argentine blue on a still summer's evening, to the inky bottomless black between ice floes out in the pack. On first visiting the Antarctic continent in 1985, I was struck by the unusual 'warm' colour of the sea under heavy overcast skies. Commonwealth Bay on this afternoon is exactly that strange deep purple-grey, like the seas I first saw at Casey. I'm yet to learn why it appears so. Perhaps because of the low sun angle, more of the warm than the cool end of the light spectrum is scattered through the clouds—a sort of prelude to a sunset.

Back at Mawson's Hut, the now dry Ted is making incredible progress with the chainsaw. He's been newly enthused by Don McIntyre and the loan of a set of waterproofs. At the apex of the hut's roof, suspended from the hip rafters, Ted has uncovered a small shelf on which scientific instruments were once fixed. The items of equipment would have included an electroscope, or electric current detector. In the cold, dry air of winter nights, the great winds cause static electricity to build up on any exposed surface. A phenomenon known as St Elmo's fire was induced, as Mawson recounts in *The Home of the Blizzard*:

> At times when the effect was strong, the night-watchman would find the edges and wire stays of the screen outlined in a fashion reminiscent of a pyrotechnic display or an electric street advertisement. The corners of boxes and the points of rocks glowed with a pale blue light. The same appeared over points on the clothing . . .

To study and record St Elmo's fire from the convenience and comfort of the hut, the AAE's young mechanic, 19-year-old Percy Correll, fixed a miniature lightning conductor above the flagpole on the apex of roof. A wire was fed into the hut and when highly charged, 'blue light radiated outwards to a distance of one inch and the air was pervaded with a strong smell of ozone'.

The overcast skies persist and the wind drops to a whisper. Cape Denison must be under the sway of a low-pressure system somewhere off the coast, pushing southwards and damming the katabatic wind. The conditions are ideal for photography of the artefacts and detritus outside the hut. With the flat, even light and an absence of shadows, the bizarre array of objects can be shown to best effect. I spend about an hour with camera on tripod photographing segments of this peculiar tapestry. My interests lie, I admit, as much in depicting the weird and the unusual juxtaposed with the ordinary, as in the serious process of recording. To me old stove lids, boots, penguin carcasses, bottles, seal flippers and sled runners border on the artistic, and like Estelle, I feel they say much about the little things, the ordinariness of domestic life.

The call has gone out for more meat with Ted particularly anxious for a steak after all his exhausting hours driving the chainsaw. Although Joan has done wonders with large pieces of 'dead thing' (the New Year's pork, for instance) the tide of opinion seems to hold that we aren't getting enough. Her vegetarian proclivities and the limited capacity of the gas stove have made her baulk at cooking steaks.

And so with the added motivation of Don McIntyre's birthday celebrations, ex-chef Dave Killick takes over for the evening and serves Scotch fillet for thirteen. Joan and the newly converted Rookie abstain, content with just the vegetables. Not put off by a pretender in her kitchen, Joan tops the meal off with a marvellous fruitcake slice.

The evening is filled with the usual child-like fun we've come to know and like from Don and Margie, complete with party masks, balloons and streamers. We all make complete fools of ourselves. Tomorrow they intend heading off for the plateau on a sledging trip—their ultimate objective Madigan's Nunatak, about twenty-five kilometres south-east of Cape Denison.

Tuesday, 6 January

About nine o'clock the wind is still blowing at 20 to 30 knots, yet Rookie confidently predicts an improving day. The barometer is up to 987.3 and rising.

As well as conducting regular radio skeds with Casey and the *Shokalskiy*, Rookie has been in frequent contact with *Spirit of Sydney*. Over breakfast, he informs us that *Spirit*'s crew has decided to abandon their attempts to get through the pack ice to reach Commonwealth Bay, and last night reluctantly turned north for Australia. They'd been trying since just after Christmas, manoeuvring back and forth through the maze, always ending in blind leads. Any thoughts of trekking off to Madigan's Nunatak with Don and Margie were first to go. In the past few days they'd just wanted to reach Antarctica. But with the summer now running short, they were left with no alternative than to turn for home.

Don is quite surprised at the difficulties both *Spirit* and the *Shokalskiy* have experienced in the pack. He'd got through without incident on each of his two previous voyages to Commonwealth Bay. The pack ice, although tricky to negotiate, had not been too heavy. It seems that nothing stays the same in Antarctica nor can anything be taken for granted. While the seasons come and go with a grinding, almost monotonous inevitability, snow, ice and sea conditions can obviously vary greatly.

The wind dies early. Rookie was right, as a superb day now looks to be in prospect. If there's ever a pattern to the weather, it seems to be in the easing of cold-air drainage, or the so-called katabatic effect, as the day progresses. It's as if the fridge door has been left open early in the morning, allowing all the cold air to drain out. With increasing air pressure off the coast and the surface layer of air warming over the plateau, the temperature inversion which 'fuels' the katabatic phenomenon is weakened and the wind correspondingly eases.

I hitch a lift to Mawson's Hut with Alan on the quad. As we come over the rise at the head of the hut valley, the sight of the Memorial Cross with its reinstated horizontal member, gives us both a lift. The valley at last looks complete. Both tragic and defiant, the cross is now a more potent symbol of the ultimate price exacted for courage and endeavour gone terribly, though not unpredictably, wrong. In balance, on the other side of the valley the hut persists, embodying the same courage and defiant strength, but with the added triumph of survival.

The Memorial Cross is becoming a regular destination for me. Once again, I climb Azimuth Hill to find Paul making adjustments to the stainless steel brackets joining the reunited cross. The rest of the 'cross restoration party' are there too. The job's quickly done and after a group photo, the others return to Mawson's Hut. I'm left to paint, revelling in the day as it slumbers on, warm and bright. Starting work, seated just below the cairn on which the cross is mounted, I think again of Ninnis and Mertz. Like all members of the AAE, they were young at a time when the world was on the brink of calamity. Had they survived, how assured were their futures? For Mertz, the Swiss lawyer, the sanctuary of his country's neutrality may have prevailed but for Ninnis, a Royal Fusiliers lieutenant, chances of survival much beyond 1914 would have been slim. In correspondence to Mawson, Ninnis's mother came to the almost sanguine conclusion that she was glad that her son had died in Antarctica, for his regiment was cut to pieces in the mud and madness of the Western Front. My painting takes in the penguin 'hill towns' along the ridge from the cross to Memorial Hill and on to Land's End. The graceful arch of the blue ice plateau rises behind. Each time I drift back from my thoughts to the reality of my surrounds, I'm made aware of the brief surge of summer life in the constant sound of penguins braying and cackling in the colonies, or calling out at sea.

While gazing across my subject, I notice the quad towing the McIntyres' sleds up the glacier to the plateau. Don and Margie have gone ahead to pitch their first camp on the way to Madigan's Nunatak and Alan has offered to take the heaviest of their gear over the first steep pitch. To my alarm, halfway up, the sleds and quad appear to slide backwards down the slope. I fear that they'll become entangled and tip over, then career back down, totally out of control. Thankfully, the quad stops and Alan is able to retreat. It looks like Don and Margie will have to do it the hard way. Later, Alan admits to me that he was concerned about sliding towards the ice-cliffs near Land's End.

After a session of landscape photography, I return to Mawson's Hut and meet up with Alan, Paul, Mike and Mal, about to set off to deploy a tide gauge for the Antarctic Division on the eastern side of Boat Harbour. Not since the work of Bob Bage, the AAE's astronomer, recorder of tides and assistant magnetician, has a record of tidal movements been made from this part of the Antarctic coastline. It's hoped that the information gathered will provide a comparison with the AAE's records. The smallest rise in tide levels is seen as important in understanding changes in ocean density, and in the Antarctic context, fluctuations in the thickness and extent of the ice shelves and sea ice. AAE data is the earliest available, and so provides a vital baseline against which contemporary results can be judged. Of particular interest to today's researchers is the issue of sea-level rise as an indicator of global warming.

Alan transports the gauge on the back of the ever-handy quad, while the rest of us walk. On the map, Boat Harbour appears like a bent arm. The gauge is to be deployed just near the crook in the arm where the fast ice is regularly inundated at high tide. As we cross to the shore, honeycombed ice crunches underfoot and occasionally gives way, its fragile crust collapsing into ankle-deep pools of salt water.

While Alan and Paul assemble the gauge, Mal dons his diver's 'dry' suit and plunges into Boat Harbour. With some difficulty, the gauge and its concrete base are manhandled into the water, then, suspended from a couple of empty water containers acting as buoys, they are manoeuvred into place by Mal. Lengths of chain and rope remain attached to the gauge to allow for retrieval at the end of our stay. Tide gauges have been deployed in recent decades in Boat Harbour, but all have been swept away by wind-blown ice. As a result, all data has been lost along with the gauges. Perhaps we'll have more luck.

Mal returns to the ice edge and grabs a video camera enclosed in a waterproof housing from Mike, then takes off to explore the harbour. He's after underwater footage of seals and penguins for the documentary. They're all out though, and Mal's soon back on the ice himself. He reports that the floor of Boat Harbour is quite lifeless, just a few tufts of algae. With repeated scouring from ice, the immediate shores of this zoo without the bars seem paradoxically bereft of life.

By 8.30 we're all back at the Sørensen for dinner, and as the evening is so calm, we enjoy the rare treat of eating *al fresco*, lounging on the rocks or legs over the edge of the verandah, enjoying a view that'd be envied by the world's most sumptuous restaurants. Commonwealth Bay shimmers, the cordon of icebergs and islets riding on a sea of softly burnished silver. The sense of well-being is palpable among all of us. Although so much of our time is spent outdoors, the chance to relax and breathe in the stillness is rare. Mawson recalls similar meals in *The Home of the Blizzard*. While building the main hut, the AAE members relaxed around what they called the Benzine Hut, a makeshift stack of fuel containers that served as the camp kitchen. Here Mawson recounts:

> Plates and mugs were seized and portions measured out, while the diners distributed themselves on odd boxes lying about on the ice.

Mawson's Hut and Transit Hut from Azimuth Hill,
Monday, 19 January.

Gouache on paper, 52 x 38cms.

ALAN GRANT, 50. Landscape architect, Brisbane. Field Leader and Australian Antarctic Division representative. Station leader, Mawson Station 1993.

GERARD (Ted) BUGG, 42. National Parks Ranger, Cradle Mountain, Tasmania. Carpenter with experience of historic timber buildings. First time in Antarctica.

GEOFF ASHLEY, 43. Conservation architect and heritage specialist, Sydney. Conservation Team Leader and Deputy Field Leader. Second visit to Commonwealth Bay.

JOAN RUSSELL, 51. Senior Public Servant, Adelaide. Cook and Camp Manager. Two previous Antarctic terms as station leader at Casey and Macquarie Island.

DAVID GILLOTT, 46. Builder, North Tamborine, Queensland. Antarctic building experience gained through a year at Casey Station and a summer at Davis.

ALLEN ROOKE, 46. Radio operator and technical specialist, Wooli, New South Wales. Communications Officer. Veteran of six winters and nine summers in Antarctica.

PAUL DELANEY, 48. Builder, Mylor, South Australia. Extensive Antarctic building experience over four winter sojourns at Mawson and Davis stations.

MIKE PIPER, 50. Film producer, Adelaide. Sound Recorder and Film Director. Experienced documentary film maker. First visit to Antarctica.

DAVID KILLICK, 29. Journalist and one-time chef, Sydney. AAP Reporter and relief cook. First visit to Antarctica.

Dr ROD GIVNEY, 41. Medical practitioner, Adelaide. Previous Antarctic medical experience gained at Casey Station in 1990.

MALCOLM LUDGATE, 49. Film maker, Hallett Cove, South Australia. Film Cameraman. Six previous summer visits to Antarctica. Award-winning wildlife documentary maker.

Dr ESTELLE LAZER, 42. Archaeologist and university lecturer, Sydney. Second expedition to Antarctica to work on Mawson's Huts.

DON and MARGIE McINTYRE Adventurers, Sydney. Sometime residents of Cape Denison.

Mawson's Hut on my first visit to Cape Denison, 12 January 1997.

Akademik Shokalskiy *pushes through pack ice.*

First landing at Boat Harbour by the AAP Mawson's Huts Work Party, 17 December 1997.

Unloading timber from the foredeck of the Akademik Shokalskiy.

We meet swirling drift and bitter southerly winds during our first reconnoitre of Cape Denison.

Evening cargo operations from the ship's anchorage in Commonwealth Bay.

Our first meal ashore, Saturday, 20 December. Geoff tucks into soup and smoked oysters on the verandah of the Sørensen Hut.

Mal and Alan manoeuvre a load of timber from the Shokalskiy*'s hold.*

Landing cargo at Boat Harbour, close to Mawson's Hut.

Dave Gillott commences excavating snow and ice from Mawson's Hut with the aid of an electric chainsaw.

Paul and helpers reattach the cross arm to the Mertz and Ninnis Memorial Cross.

Moving the Apple Hut, Saturday, 27 December.

The extensions to the Sørensen Hut near completion.

Painting on the Edge of Midnight. Wednesday, 7 January.

Gouache on paper, 52 x 38 cms.

Penguin 'Hill Towns'. West from Azimuth Hill, Sunday, 4 January.

Gouache on paper, 38 x 52 cms.

Snow Clouds. East across Commonwealth Bay, Sunday, 25 January.

Gouache on paper, 38 x 52 cms.

Adélie penguins and Weddell seals on the northern shores of Boat Harbour.

Nesting snow petrels, near Azimuth Hill.

Adélie penguin feeding its chick, John o'Groats.

Weddell seal on honeycombed ice, near the tide gauge site.

Weddell seals near Boat Harbour.

Evening view to the east from the Sørensen Hut.

Adélie penguin colony below the Sørensen camp, the day of the storm out to sea, Friday, 23 January.

Dave Gillott and Ted fixing new pine boards to the eastern face of the workshop roof.

Dave Gillott and Ted recladding the workshop roof of Mawson's Hut, while Paul cuts boards below.

The Transit Hut on a perfect summer's evening.

The author and Mal at John o'Groats, Thursday, 1 January. (Photo by Mike Piper)

Snow clouds over Commonwealth Bay. Gadget Hut is in the middle distance.

Dave Killick and Paul collecting water in a blizzard.

The author negotiating a path from the tent to the Sørensen Hut.

(Photo by Geoff Ashley)

Mike and Mal film a piece to camera, 'Hi, I'm Geoff Ashley, welcome to Antarctica!' Saturday, 17 January.

Between Ice and Tide.

Friday, 16 January.

Gouache on paper, 52 x 38 cms.

The Cliffs of Land's End.

Wednesday, 28 January.

Gouache on paper, 52 x 38 cms.

A blizzard approaches.

Mal filming Alan running survey levels from the AAE's bench mark, Thursday, 15 January.

Joan goes 'shopping'.

Paul working on repairs to the ceiling platform inside Mawson's Hut.

The northern end of Mawson's Hut after a blizzard.

Stable door to the Magnetograph House.

The weathered face of the western verandah of Mawson's Hut.

Inside the Magnetograph House. A pile of AAE clothing lies in the right-hand corner.

Rod and the author 'walking the dog', Thursday, 22 January. (Photo by Alan Grant)

The remains of the AAE sledging dog, nicknamed 'Rex'.

Alan says a few words before we devour Joan's steaming Australia Day pudding.

Rookie at work in the communications area (entry vestibule) of the Sørensen Hut.

'Brother' Ted relaxes outside the Granholm Hut, otherwise known as the 'Lodge' of the 'United Brotherhood of Carpenters and Joiners of Antarctica' (our three builders).

New Year's dinner.

Alan digging out the dislodged door of the Absolute Magnetic Hut.

A shelf in the darkroom.

Looking towards the library shelves in the living section of Mawson's Hut, showing the repaired framing to the ceiling platform.

The 'Brothers' pose on the roof of 'Doug's Shack'. Photo session preparatory to Australia Day, Sunday, 25 January.

Advertising leaflet found in the darkroom.

Examples of the tins left on the library shelves.

Print above Mawson's bunk.

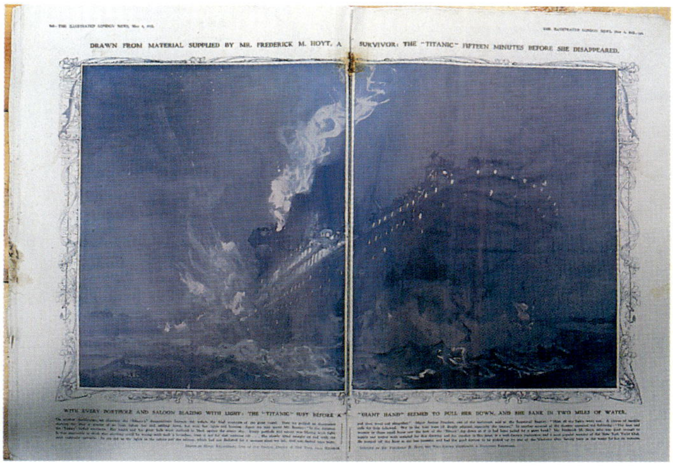

The Titanic *disaster as portrayed in* The Illustrated London News *of 4 May 1912.*

The Shokalskiy *on the homeward voyage, Thursday, 5 February.*

Daylight streams into Mawson's room for the first time in more than 84 years.

> Many who were accustomed to restaurants built tables of kerosene cases and dined *al fresco* . . . We can all paint our own picture of the good times around the Benzine Hut. We worked hard, ate heartily and enjoyed life.

Mawson's words are as true today as they were in 1912.

After dinner, I return to paint at the Memorial Cross, yet again. This time it's the view to the north that captivates my interest. In the late evening the frozen canopies of the nearby Mackellars are bathed in an ever-strengthening gold. A floating mosaic of brash ice has drifted around the point and it too glows with the last of the day's alchemy. I work quickly as the midnight cold deepens.

All the near islets are capped by fanciful turrets of ice, yet in the outer cordon, a number are low and bare. The windward islets have no protection and so are often deluged by spray breaking right over them. The spray freezes and accumulates, as Mawson put it, in 'monstrous cakes'. These cakes fluctuate in height between summer and winter, and on the smaller rocks, the assaults of the most ferocious storms can blast the ice clean away. But throughout, the inner chain acts as a permanent windbreak to the outer, and the seaward islets are therefore bereft of ice. Allowing easy access to the sea, this precious ice-free land teams with wildlife. Mawson estimated that 150,000 Adélie penguins crowded the largest of the outer Mackellars alone. So close and yet unattainable, unfortunately we won't get to see any of Mackellars at close hand unless there's a sustained lull in the weather when the *Shokalskiy* returns—something none of us expects.

As the sun drops below the plateau, the cold intensifies. My eyes and nose stream and my fingers are numb and stiff. I stop painting about midnight, but decide to stay out and enjoy the glories of the night for as long as possible. While I can no longer paint, I can certainly take photographs.

Before moving off the ridge, I pause to photograph the Memorial Cross silhouetted against a sky now flaring with pinks and yellows. Dark against streams of cloud pointing inland to tragedies past, the cross seems to assume a heightened and even more poignant theatricality. The dazzling light of day has long since receded to pastel, shadowless blues, but above, the sky continues to shift in a palette of infinite variety: from gentle mauves, yellows and even green, to bolts of orange fire.

Down on the flats, I crunch across the ice to a deserted Mawson's Hut, and for the first time, I think I grasp that rare sense of solitude from being, momentarily, quite alone in Antarctica. It's a heady and almost risky thrill, which naturally leads me to think of Alfred and his year alone at Cape Denison. As much as I savour this special moment, I know that I wouldn't want to spend that much time alone, even surrounded by so much sublime beauty.

At three o'clock I'm too tired and cold to continue, so reluctantly begin the trudge back to the Sørensen. The sea ice near Boat Harbour is almost deserted. Only a handful of seals slumber on. Weddells are generally nocturnal and prefer to go sea and feed at times of minimum light. Despite my growing eagerness for a warm sleeping bag, I can't resist photographing the Weddells under a seaward sky grading from purple, through mauve, violet and pink to pastel primrose. I arrive back at the Sørensen, exhausted but elated. All's quiet. The usual late-nighters are fast asleep.

Wednesday, 7 January

I sleep until 10 o'clock and get up feeling heavy. It looks like another good day with the wind true to pattern, tapering off as midday approaches. Over breakfast, I write a couple of e-mails which Rookie promptly manages to lose somewhere in the ether. However, it's not our redoubtable operator's fault, just computers! After quickly retyping my messages, I head off for Mawson's Hut, arriving in time for lunch.

The day continues to improve, so Alan and I agree that it's a good time to start one of our more fiddly tasks—a complete terrestrial photogrammetric survey of the hut's exterior. Our first challenge is to assemble the bulky stereoscopic camera lent to us for this purpose by the Hydro Electric Corporation in Tasmania. Keith Stove, the enthusiastic head of the Hydro's survey branch, gave us a quick explanation of its features back in Hobart. In theory, we should know how everything fits together.

We unpack the camera and its tripod from two formidable plywood boxes and after a considerable amount of head scratching and a couple of ruined glass plates, feel confident that we know enough about the 'beast' to continue. The twin cameras and sturdy connecting arm are heavy enough, but when mounted on the massive tripod, Keith's pride and joy can only easily be manoeuvred by two people. Lugging it from point to point around the building proves to be something of a chore, particularly when I jam a finger between the tripod's heavy, collapsing legs. It looks like technology not far advanced past that of Mawson's era, but the superb, precision optics of the two Carl Zeiss lenses will provide stereoscopic images without distortion or aberrations of any kind. The production of exact scale drawings will be possible from our stereoscopic photographs of each façade of the hut.

I repeat the photography of the workshop using my own medium-format camera and tripod. If my images overlap sufficiently, what the stereoscopic camera does optically will also be achieved by computer analysis. John Osborn, a researcher at the University of Tasmania's Department of Geography and Environmental Studies is interested in putting my images to the test. In discussions with John just before we left for Antarctica, he described the relationship of the computer to the camera lens to be analogous with that of the human brain and the eye—each corrects and enhances information from an imperfect optical instrument. It'll be interesting to compare the images that result from the two cameras.

Mawson's Hut is well on the way to becoming the most photographed and measured building in Antarctica. It's hard to even think of a more photographed Australian building—the Sydney Opera House, perhaps? I jest, but then I've often thought about Mawson's reaction to all the attention his winter quarters are receiving. The rational scientist might consider the money and effort better expended in research, yet the proud but humble explorer would probably have been more than a little flattered.

With the photogrammetric recording of the workshop complete, the roof can now be reclad. Following Geoff and Paul's investigations of a few days ago, the decision has been made to fix the new cladding directly over the old. Ted, Dave Gillott and Paul take advantage of the calm conditions and work on well into the evening, quickly fixing boards over about half the western face of the workshop roof. Even and neat, the new honey-coloured roof creates quite a contrast with the old timbers and looks strangely out of place. In an instant, the patina of 86 years is gone. I suppose a few years of Cape Denison blizzards will help make the new boards look 'at home'. Joan sends soup and bread over to the Weatherhaven at around 7.30 and we all then work through until about 8.30. The western face of the workshop roof is now complete.

Geoff and I walk back to the Sørensen, detouring to the rocks just to the north of the Absolute Magnetic Hut, one of the three small observatories built by the AAE. A significant aspect of the AAE's scientific work concerned studies aimed at detecting minute changes in the earth's shifting magnetic field. Cape Denison in 1912 was several hundred kilometres north-west of the South Magnetic Pole, yet today the restless pole lies out to sea near Dumont d'Urville. By continuously monitoring the magnetic field, the rate and direction of movement and ultimately the reasons for change could be determined. Although the cape was not as close to the pole as to be considered ideal by the AAE's magnetician, Eric Webb, readings were taken from a number of stations during the entire eighteen months of the expedition. The AAE's work complemented and continued that of Mawson, David and Mackay from their 1908 journey to the Magnetic Pole.

The Absolute Hut was built as a base station in which a dip-circle needle and tripod and a magnetometer were housed. All other magnetic observations were periodically calibrated against readings obtained on the Absolute Hut's instruments.

However, it's not the hut that Geoff has brought me to look at. About ten metres from the ruined building, he points out a rusty tin wedged in the rocks and cemented fast by ice. Such items are almost commonplace in parts of Cape Denison, but even a casual glance reveals that this item is somewhat different from the usual AAE detritus. The tin is actually a home-made lantern fashioned by Webb. It was used by him in winter sights from the Absolute Hut to nearby survey marks and is referred to in the *AAE Scientific Reports* published in 1925. With hinged door, copper-wire handle and latch, soldered joints and carefully fashioned vent and cowl, the lantern is quite craftsman-like in its construction. Despite it having survived more than 80 years of exposure, with most of its features intact, Geoff is worried about its continued prospects.

One corner and a large part of the back have rusted away and the whole lantern will soon be in danger of fragmenting. Geoff is keen that it be repatriated to Australia, for conservation and eventual display.

Yet another night for painting, this time just below the Sørensen. The extravagant skies over these last few evenings have been a great source of inspiration. I race against the cold, quickly describing the vibrant pink and orange curtain behind the Mackellars. With the aid of some antifreeze in the water, it's done—a quick but satisfying study. A couple of beers with the usual stayers—Dave Killick, Ted, Alan and Paul—then bed around 1.30.

Thursday, 8 January

Another still night. These conditions are almost bordering on the monotonous. At nine o'clock the wind—if that's not too strong a term—is a delightful 8 knots and the barometer already 994 and rising. Overnight the temperature dropped to −2.5° Celsius.

I try to make an early start from the Sørensen, but am trapped by Mike and Mal shooting a mock sequence of Alan Grant having a phone conversation with Rob Easther back in Hobart. The action takes up all of Rookie's communications area, blocking any entry or exit from the hut. Mike and Mal are good fellows, and they have their job to do like the rest of us, but the camera's constant presence is becoming quite intrusive for some. Outdoors it can be particularly trying. The search for the best camera angle often necessitates several takes in less than pleasant conditions and is increasingly the cause of a few grumbles.

At last away, I head out into another one of those Antarctic days where superlatives are inadequate. The air is warm and pleasant and the snow melt continues apace. Boulder fields widen and the expanse of blue ice grows. Under more defined gullies, the subterranean trickle of meltwater can be heard.

As usual, the Weddell seals provide a diversion on the way to Mawson's Hut. This time, I'm distracted by the song of these sirens of the ice. Several animals call back and forth in a changing repertoire of clicks, chirps and high-pitched squeals reminiscent of whale song. Perhaps it's the fine day that's making them so vocal, but until now I've been oblivious to their calls. In the water these sounds are the Weddell's means of communication in the dim and often dark world under the ice. They may also be used in navigation and the location of prey. Reasons or theories aside, I could linger for hours just listening to their strangely enchanting calls, but work awaits.

At the hut, Ted, Dave Gillott and Paul are screwing down the new Baltic pine boards that were tack-nailed in place the previous day. Ted seems much happier working on the roof, savouring the warmth after his stint behind the chainsaw. The damp and cold of the Cradle Mountain–Lake St Clair National Park are one thing, but for Ted being doubled over or stretched into awkward corners with a whirring chainsaw in hand has been a draining and at times miserable experience. Mainly through his efforts, an estimated sixteen cubic metres of ice and snow has now been removed. Out in front of the hut, Estelle also enjoys the sunshine and calm as she painstakingly starts to map the artefact field.

The day continues to warm as the wind drops to a whisper. I quickly finish the photographs of the workshop for John Osborn at the University of Tasmania before the remainder of its roof is reclad. I'll return to photograph the living section of the hut later—the weather is too good not to be painting.

Despite Land's End being only a five-minute walk from Mawson's Hut, it's a corner of Cape Denison I've yet to explore. The gradual trudge up the glacier to the edge of the ice cliffs reveals a vastly different view of the hut valley with Commonwealth Bay beyond from any I've become familiar with so far. Islets and icebergs, ride like ships at anchor on an oily sea, while down the valley, the roof of the hut is barely visible over a protective brow of rock. From there, tracing an arc to the south, the steepening curve of the glacier sweeps upwards to the blue ice of the plateau, broken only by two piles of enormous boulders.

Arriving at Land's End, the disjunction between the security of a rocky foothold and the wilderness of tumbling ice is even more abrupt than at John o'Groats. Only about fifty metres from the end of the ridge running south from Azimuth Hill, across a narrow inlet, a huge scoured wall of ice abuts the last of the bare rock. Riven by cavities and streaked by dirty-brown bands, the ice wall is like a face from some nightmarish anatomy text, with membranes, sockets and sinews all exposed. The dirty ice interleaves layers of entrapped rock, ground from the underlying strata somewhere deep beneath the plateau. About to be disgorged into the sea, these folds, like lines on a furrowed brow, tell of interminable stresses within the ever-moving ice.

Painting in such gentle conditions is a delight. Totally without need of shelter, my bare face and hands bathe in the warmth of the afternoon sun as it traces a shallow arc towards the plateau. At a glance the sun appears motionless, its passage marked only by deepening folds in the ice. But as the shadows become bolder, my ability to define space and form is greatly tested. Added to the changing forms, I'm forced to grapple with countless shades of blue from the lightest turquoise to the most intense indigo.

As at John o'Groats, there are large numbers of penguins here. But rather than nesting on roughly benched terraces, their colonies cling to a near-vertical slope. The penguins near me are a raucous bunch, constantly braying and calling and occasionally fighting. The calls are all part of the penguin's ritualised behaviour: from the ecstatic, flipper-beating display of the males, to the bonding exhibition of paired birds. In the latter, each bird fronts its mate with neck stretched and both emit a throaty cackle rising to a sharp crescendo. The chicks are growing rapidly. Many are now quite large and have reached about two-thirds the height of the adults, but in looks, may as well be a different species. The first moult for these fluffy, grey, and roughly cone-shaped birds with pink bill and feet is still a few weeks off.

As I paint, I become aware of the quad's motor labouring heavily, somewhere in the distance. It sounds like it's heading for the plateau. Alan must have decided to pay another visit to the McIntyres. A couple of days ago, he and Don located the mummified remains of an AAE dog about five kilometres south of Cape Denison. The dog was first spotted by Alfred on one of his lone ski trips but no map reference noted. Alan retrieved one of the animal's front paws and brought it back to camp, principally to show Estelle. I don't think she was too pleased to receive this gift, as the arrival of the dog in our lives potentially means more work and some complex decisions.

Returning to Mawson's Hut, I'm surprised to learn that Alan's still not back from the plateau. Mal's up there too. He went along to film and to help Alan document the find before the Antarctic Division is asked for a decision on the dog's fate.

Later, over dinner at the Sørensen, Alan relates a few details from a rather eventful afternoon. Despite the perfect conditions, he and Mal got lost on the way up to their rendezvous with Don. They strayed onto the edge of a crevasse field, where the front wheels of the quad broke through a snow bridge and teetered on the edge of a rather deep hole. Alan freely admits that they were both a little scared. But they managed to extricate themselves and having regained their bearings, proceeded inland, a calamity narrowly averted.

Again graced by a calm evening, I decide to return to Land's End and produce another variation on the afternoon's study. Arriving around ten o'clock, I quickly set to work on a subject utterly changed in the space of a few hours. As shadows drape over the horizon, positive turns to negative and the once dazzling white line of the plateau sits deep and cold against a yellowing sky. Without the play of direct sun and shadow, the ice wall flattens, its deep cavities and sockets growing ever darker and more portentous.

My composure is finally whisked away on the chill breeze of midnight. I pack up quickly, eager for warmth. But watching the sun at last brush the horizon, I feel compelled to stay out for a while and photograph the evening sky. I walk across the glacier to the highest pile of boulders south of Mawson's Hut and there wait for the colours of the night to intensify. But they never come. Instead, cloud begins to move in rapidly from the south, so I take the easy choice of the Sørensen and the thought of a hot drink.

As Estelle is still out mapping artefacts in front of Mawson's Hut, I wander down to see her on my way home. Rod's there too, concerned that she's out late in the growing cold. She's easily persuaded to stop and so the three of us are soon crossing the last ridge before the Sørensen. I walk ahead, and just before the final boulder field, become aware that I'm alone. Looking back, I see Estelle crouched on the snow holding her right arm, with Rod bent over her. Rod calls to me, indicating that she'd just slipped over on the ice and was all right—something we're all doing occasionally. I continue on to the hut. Rod and Estelle follow a couple of minutes later, but instead of relaxing for a while before bed, he ushers me out of the kitchen and proceeds to examine her. On the way out she protests to me that her arm is fine, so I think nothing more of it and head for the tent. In the meantime a solicitous Joan arrives, up out of her bed, sensing that something's wrong. Her face tells the true story.

Friday, 9 January

Our string of fine days inevitably comes to an end. It's blowing hard and a heavy blanket of grey hangs over Cape Denison.

Before I can make plans for the day and pack my gear, Rod takes over the dining area to relocate what has turned out to be Estelle's dislocated shoulder. He needs an operating table. I'm now feeling a bit guilty over my half-hearted concern for her last night. I had no idea of the severity of the injury. For privacy, our Australian flag is draped over the doorway and all but Joan and the patient evicted from the make-shift surgery.

Hastily grabbing a camera, I decide that it might be best to get right away. I head out into a very different day from yesterday's shimmering brightness. An arc of gun-metal grey cloud cuts a swath across the sea, lightened only by the glow of the ice blink over Commonwealth Bay's outstretched arms. I head towards the Absolute Magnetic Hut and photograph the shattered ruins against a wild and threatening sky. Of all the dramatic skies we've witnessed so far, none has looked more menacing.

Returning to the Sørensen, I find the operation successfully completed, with Estelle drugged and sleeping on the lower kitchen bunk. Snow begins to fall ahead of a quickening gale. It certainly doesn't look like a painting day, so I head for Mawson's Hut around noon. Due to the weather, there's no work happening outside. All efforts are directed to continuing the excavation around the collar ties and ceiling platform in the living section. Geoff decides that the precipice of ice, snow and hoarfrost hanging over Mawson's cubicle should be removed. While it's technically outside the scope of the work we've been permitted to do, Geoff feels that this five-tonne mass would soon fall and do significant damage to the shelves, bunk and artefacts beneath.

The radio message comes through from Joan that lunch is on its way. Outside, the snow flurries of a couple of hours before have turned into our first real blizzard, so we all bundle hurriedly into the sanctuary of the Weatherhaven and there keenly await the arrival of 'Killick, the Human Pie-cart'—as I've recently dubbed him. A rider of more conventional bikes back home, Dave has taken to the quad, and is often seen astride this fun machine, tearing off across the snow at any excuse. One of his self-appointed jobs has been that of 'meals on wheels' driver. It quite suits his work, as he often needs to be in two places at once: both at the Sørensen, writing and sending stories; and at Mawson's Hut gathering digital pictures and monitoring progress. If he can bring the lunches over or run errands back and forth on the quad, then so much the better.

We're all relieved when at last we hear the muffled sound of the quad churning through the deepening drift outside. A minute or so later the tent's zip parts and in steps a breathless Dave, covered in snow and carrying a large cardboard box and a backpack. The pack contains the usual thermos flasks brimming with hot soup, while the box holds a special treat—three delicious pizzas sporting individual toppings, all made by the 'human pie-cart' himself. Dave had left the Sørensen in reasonable visibility, but nearly lost his way around the top of Round Lake, only regaining his bearings when suddenly confronted by one of the large boulder piles looming out of the blizzard. The Weatherhaven is soon steaming, as the snow on our ventiles melts and nine or ten weary faces guzzle hot soup and munch pizza. Outside, the blizzard rises, drift-filled gusts sounding like sand flung against the walls of the tent.

We work on in Mawson's Hut until the late afternoon, occasionally checking the conditions across the valley through the skylight that opens onto Ted's chute. About 6.30, the Memorial Cross can only just be made out. It's time to go. Everybody gets back to the Sørensen safely with Mike and Mal filming the return. Soon after our arrival back at camp, the visibility drops to less than fifty metres.

Before we're totally marooned, Paul, Dave Killick and I head to Alga Lake for water. The walk up is decidedly unpleasant, as we cower into the stinging drift and a 60-knot wind. Catching my bearings, I look up briefly midway to the lake to see 50-metre-high flurries of drift, like great swirling plumes of smoke, driving down from the plateau and on in an instant past our camp. At the lake we engage in the usual struggle with saucepan and water bottles, while around our feet snow swishes past, brushed from the hard smooth surface of the ice in a fine unsullied powder.

Back at the Sørensen, the verandah and our outdoor larder are fast drifting up with fine sticky snow. Our feet are covered in it, so the accepted etiquette is a quick brush if the broom's handy or a kick with each foot against the sill before opening the door. In such conditions, indoor conversations are punctuated with the repeated sound of boots rapping on the hut.

Inside, the humidity climbs as a row of snow-caked ventile jackets and overalls become sodden, then steam. Condensation starts to trickle down the windows and the occasional drip falls from the ceiling. Perfect for the cold and dry conditions that might be experienced at altitude, or in winter, the finely woven double-layered cotton ventiles are not suited to the comparatively 'warm' blizzard raging outside. Wet snow sticks easily to the fine fabric, then melts quickly with the body's heat. Wet clothes and Antarctica would be a potentially dangerous combination. Perhaps the Antarctic Division should reassess the clothes it issues—some designs and styles haven't changed in 50 years.

Mike and Mal are still out there in the blizzard, capturing some of the local 'colour' for the documentary. The worse the conditions are the more Mal gets involved with the drama. Professional, energetic and stubborn, Mal is in many ways something of a latter-day Hurley. He'll go to enormous lengths to get the most dramatic and evocative shots possible, often working in considerable discomfort and at some risk. With a television monitor on hand, we've been able to glimpse some of the material that Mike and Mal are accumulating. To the untrained eye, it's looking extremely good. Mal the perfectionist may be a little more reserved in his judgement.

As with so much of our equipment, I can't help noting the contrast between Mike and Mal's gear and that used by the AAE. While Mal cossets and fusses over the complex computer systems and electronics of his Sony cameras, charging and recharging batteries daily and desperately keeping snow out of their delicate workings, Hurley struggled with an ungainly hand-cranked cinematographic camera weighing, with tripod, nearly forty kilograms. In his book *Argonauts of the South*, Hurley described his battles in capturing the trials of daily life in 'moving pictures':

> when the wind attained a velocity of 120 miles an hour, I was lifted bodily, carried some fifteen yards . . . and dumped on the rocks. I was reduced to crawling on all fours, and finally had to call an assistant [Webb], who followed close on my heels, dragging the tripod, while I went ahead with the camera.

Should I be out there too, capturing the blizzard in still photographs? I let the question go unanswered in my mind. Anyway, it's Rod's birthday tonight. Joan turns on an excellent spaghetti bolognaise with an iced fruitcake to follow. She's a little peeved that there are two missing, partly out of concern for their welfare but also because they're not back for dinner.

Our intrepid duo finally return about midnight. Cold, covered in snow and quite flustered, they had been forced to hole up in the Granholm for a time when the visibility got too low. Hardly shed of his ventile jacket, Mal riffles his gear for a soft brush, then proceeds to fuss over his precious camera, carefully whisking away snow before it melts and seeps inside.

Saturday, 10 January

Waking several times in the night feeling cold down one side, I soon realise that snow has drifted up in the lee of the tent. It pushes inwards against my body. There's not much I can do about it until morning, so I wriggle to the inner edge of my very narrow stretcher. Hanging half over the edge of the stretcher is as uncomfortable as lying against the tent wall is cold.

I turn out of the tent to a barely visible Commonwealth Bay and still lots of driven snow. Rookie's situation report states, 'Weather still being obnoxious . . . Temperature staying below zero now. Barometer undecided on rise or fall?'

After a prolonged breakfast, the visibility seems to be improving, so a few of us head for Mawson's Hut. The walk over is not too difficult, despite a 40-knot wind and deep banks of snow. Fine drift blows like sand across the ridge above Gadget Hut and in rocky areas, a footing is hard to find under an airy cover of dry powder.

In the hut valley, the drift's thicker and the wind gusts up to ten knots faster. Even within the few hundred metres separating the Sørensen and Mawson's Hut, there's a marked difference in the severity of the Cape Denison climate. If the cape is the windiest place on the planet, then Mawson's Hut must be at its epicentre. The spectral forms of the Transit Hut and then the main hut itself drift in and out of view through the veil of blowing powder. The savagery of this place in Mawson's time as now, makes the ghost-like hut a true symbol of endurance.

By the time I round the corner of the workshop it's nearly impossible to see past the next footstep. Then, just before the Weatherhaven, I stumble and fall face-first into a long tail of drift that's built up in its lee.

Soon after the hut was finished in February 1912, snow drifted up around it to the point where only the roof was visible. Mawson notes:

> The snow continued to descend in torrents and, but for the wind, the Hut would have been quite lost to sight. The packed drifts rose higher round the walls, and back eddies brought the snow past the canvas flap at the entrance, though situated on the lee side, until the veranda became choked. Frequent shovelling was necessary to maintain freedom of exit.

In winter, a network of tunnels that Mawson likened to the 'Labyrinth of Minos' provided access to and from the hut. Through the darkness of a winter blizzard even the roof became hard to find. Those returning to the hut would feel their way, groping for clues, often ending up on top of the roof itself.

> In very thick drifts one might stand on snow-covered portions of the roof without any notion of the fact, hearing nothing of the noise below however boisterous, oblivious to the proximity of refuge and even despairing of ever reaching it.

Work for today is again confined to the interior of the hut. Excavation on the ceiling platform is nearing completion with numerous artefacts now emerging from the ice. The chainsaw has been discarded for a couple of pinch bars. Geoff maps the platform, noting the position of the emerging finds, while I photograph them, still in situ. It's hard to decipher much of the jumble but there's some scientific equipment there as well as old tins, glass and scraps of clothing.

For a break, I decide I should emulate Mal and photograph the blizzard raging unabated outside. I struggle over to the Granholm, then to the penguins nearby. They seem to be having a miserable time of it, adults and chicks alike caked in snow and hunkered down against the gale. The chicks, desperate for warmth and protection, would burrow right inside their parent if they could. Some chicks are starting to group into small crèches. They flounder about, climbing over one another in search of greater sanctuary from the shifting group while others stand motionless, almost stoically resigned to the blast. Blizzards are heavy killers of the young chicks. Once the snow on their down melts, effective insulation is gone. Many Cape Denison chicks will freeze to death in this and a likely string of blizzards before the colonies disperse in March.

Out to sea, swollen grey clouds scud away from the coast and between fresh flurries of snow, scraps of blue appear in the sky.

Returning to Mawson's Hut I pause to take some quick photographs of the work site in the midst of the blizzard. All I can make out is the vague shape of the Weatherhaven and the hut behind. Our building materials are buried under great banks of drift.

I arrive back at the tent to find Alan there with the lunches. He'd made an intrepid attempt to get over on the quad but was beaten by the drift within 30 metres of the work site. Ted and I stagger up with a couple of shovels and help him dig out the well-bogged quad, barely visible above the snow. After a few attempts, rocking the quad back and forth to get some traction, it finally grabs, bursting out of the bank and spraying snow over the pair of us. Moving about is getting difficult. The drift is very deep and soft in places making an even or predictable footing impossible.

A number of us are now wet and cold from all the blowing snow, so we decide to finish early and head for the Sørensen. The quad has drifted up again and won't start, so Alan manoeuvres it around to face into the wind and abandons it for the night. We arrive at the Sørensen, very thankful for a warm hut. Walking in the blizzard, or more accurately stumbling, lost its novel appeal for me some days ago.

After a tasty meal of meat loaf and vegetables, we embark on what at times seems an interminably long meeting. Discussion centres on a review of progress both on the conservation program and consolidation of the Sørensen camp. Alan is keen to see the Granholm dismantled and moved near the Sørensen in a similar manner to the Apple.

Then there are extensions to the Apple itself—breaking it apart and inserting the additional prefabricated panels brought down on the *Shokalskiy* specifically for the job.

With an additional building plus the extended Apple (Melon) in place around the Sørensen, a very commodious little camp would be created and the 'visual catchment' of the hut valley, as Geoff calls it, would be substantially tidied and improved. In the short term, most of us could then vacate the tents. And for ANARE and any future AAP Mawson's Huts Foundation work parties, a properly functioning summer station would await their arrival.

While Geoff supports the logic in all these plans, he's getting quite anxious about the progress of the conservation program. It's going well, Geoff reports, but he's still concerned that diverting our energies away from the historic huts as January advances and the weather deteriorates will only leave us with a list of incomplete tasks at the end of our stay.

Geoff seems quite tense. To ease his anxiety, Alan finally agrees to suspend home improvements for a week and then reassess progress on the conservation plan. Joan weighs in with her motherly concerns for our safety and her worries about people coming home tired each day. She feels we should continue some of our home-making tasks in the interest of our collective well-being. I still feel, though I don't say, that if there were a little less early morning drinking, some people might not be so tired.

Other items discussed don't help Geoff's demeanour. Dave Killick has been filing stories, as any journalist might, without having them vetted by Geoff. Through a misinterpretation of some technical detail, a recent story was run in the press that didn't reflect Geoff's intentions in relation to recladding the workshop roof. A spirited, though civil discussion ensues, pitting journalistic independence against the need for technical accuracy. Dave stands his ground but agrees to be more mindful of technical detail. He also offers to circulate his stories—after they've been sent.

The beer supply ran out over dinner, so around 10.30 Dave Killick and Ted do the valiant thing. Braving the snow, they head for the stockpile over at the worksite, returning half an hour later with a few dozen stubbies in ruck sacks. One beer and I'm out in the snow myself, but I venture only as far as a warm sleeping bag. Since it's Sunday tomorrow, I suspect a few of the others will be some time.

Sunday, 11 January

The blizzard is still blowing, but appears to be waning. Paul's not convinced. The longest blizzard he's ever endured lasted for a week.

It's a day to take stock: to write, eat, chat, read and drink coffee. The expedition library is not extensive, consisting mainly of technical reports centred on Geoff and Estelle's work, but the hut does have its own permanent copy of *Mawson's Antarctic Diaries*. Edited in a thorough and scholarly manner by Fred and Eleanor Jacka, the diaries are Mawson's first-hand record of his three Antarctic expeditions. They make fascinating reading.

The first and most obvious question for a diary keeper is 'What were the AAE doing on this day?'. Unfortunately there's no entry for 11 January 1912 in the abbreviated log that Mawson kept during the early days at Cape Denison. The desperate rush to get ashore and establish winter quarters must have precluded all else. In contrast, the entry for the same day in 1913 tells of dramatic and terrible events. This was the first day that Mawson moved on alone after burying Xavier Mertz. About 160 kilometres south-east of the hut, he writes:

> ... 5.30 p.m. I felt nerve-worn ... and had to camp—so worn that, had [it] not been delightful evening, could not have found the strength to erect tent. If providence can give me 20 days weather like this ... surely I can reach succour. My whole body is apparently rotting from want of proper nourishment.

It was not until 8 February that Mawson staggered back to Cape Denison.

Even during the worst of the awful ordeal that followed the deaths of Mertz and Ninnis, Mawson writes in a spare and generally unemotional style, not drawn to romanticism or introspection. His diaries throughout are essentially the writings of a rigorous inquiring mind, a scientist. But for the student of human behaviour under extreme or adverse circumstances they sometimes provide glimpses of the 'real' AAE and not just the 'happy band of brothers' that Mawson is want to portray in *The Home of the Blizzard*.

Any group is bound to comprise individuals of varying talents, capacities and attitudes. While the young men of the AAE were on the whole an enthusiastic and idealistic team that pulled well together, Mawson notes that a few individuals didn't always match his own aspirations or expectations.

On 3 May 1913, ropes connected to the wireless masts were noticed to be badly chafed and in need of urgent attention. Mertz and Ninnis rushed to the assistance of Walter Hannam the radio operator, but others were less enthusiastic. Mawson notes:

> I was sorry to see that the others, who were not engaged on special work, excused themselves and one even refused to go. With all due consideration for the frailty of such individuals, this is scarcely a brotherly feeling, to say the least of it.

Leslie Whetter comes in for particular and repeated mention for only ever begrudgingly doing the minimum when it came to his allotted tasks. The 29-year-old New Zealand surgeon was supposed to have been medical officer for one of the three bases that Mawson initially planned to establish. When the smallest of the groups was amalgamated with Mawson's own party at Cape Denison, Whetter was made subordinate to the younger and less experienced Archie McLean. The mutual dislike between Mawson and Whetter is clear from Mawson's writings. Mawson would often call Whetter to his room and there reprimand him. All this must have been within earshot of the others; there was no privacy in the tiny hut.

During the second winter, Mawson had a more perplexing personnel problem to manage. The two or three recalcitrants were gone on the *Aurora*, but the small party that remained had to cope with a severe mental disturbance in one of its members. Sydney Jeffryes, a radio operator from the *Aurora* was the only addition to a core group of five that remained behind in hope of the return of Mawson's sledging party. He had previously been rejected by Mawson as a candidate for the expedition, but somehow inveigled himself south as Hannam's replacement.

As Mawson's own health gradually improved, Jeffryes' condition deteriorated. In his worst delusions he imagined the others were plotting against him and sent radio messages to this effect to Macquarie Island. Mawson notes on 21 September 1913 that Jeffryes 'Says I have him under an hypnotic spell, thanks to which his life has been saved from murderous intentions of Bage, Dad [McLean], Madigan and Bickerton'. He rarely ventured from the hut and was eventually relieved of his duties by Mawson. For the seriously debilitated Mawson, the strain of managing Jeffryes was almost too much to bear at times. Nothing more was mentioned of him after the *Aurora*'s return to Cape Denison in December 1913. Tellingly, Jeffryes is absent from Hurley's photograph of the Cape Denison and Macquarie Island parties, taken shortly after all were back on board the *Aurora*.

As we sit hut-bound for the day, it's interesting to reflect on parallels between ourselves and the AAE. More than eighty years of social change mean some values and attitudes have altered significantly, particularly in regard to leadership and authority. But we still face many of the same difficulties and privations, albeit for a much shorter time and with far better contact with the outside world. Above all, we must remain self-motivated, just as most the AAE were, for our goals to be fully realised. Small cracks have begun to appear in our own society. Would they widen into crevasses with an enforced or prolonged stay?

The day drifts by. At times I find myself half-dozing as I read. With the strenuous demands of living, working and just existing here, plus the lack of restful sleep, fatigue is starting to catch up with most of us. It's only on days like this when a complete stop is enforced, that one realises how wearing our Antarctic sojourn is becoming.

I spend a couple of hours in the afternoon writing some e-mails, then, as it's Sunday and cook's day off, Geoff and I volunteer to prepare dinner. Freeze-dried beef curry seems to be the consensus, so first we need to make a 'shopping trip' to the food boxes out on the verandah. Fortunately, the boxes aren't drifted up and the correct one is readily at hand. This, I hasten to add, is not always the case and shopping can be a frustrating and finger-chilling task.

Freeze-dried beef curry is one of a number of varieties in our larder. They're all identical to those eaten by the Australian Army in the field and are generally very good. I do however suspect the freeze-drieds of being at least partly responsible for the heartburn epidemic that's still raging out of control. So persistent has the epidemic been among some of us, that the small knoll on which the Sørensen sits has come to be known as 'Heartburn Hill'.

The basic constituent of dinner comes in thirteen individual foil packets. It's pretty plain fare simply reconstituted in water, so Geoff and I tart it up with onions, garlic, ginger and cashews. Served on the traditional bed of rice—delicious.

Monday, 12 January

As I'd forgotten to clear the drift from my side of the tent, I again suffer from a cold back and consequently don't sleep well. I'm not sure whether it's my imagination or my oversensitive balding head, but I think the nights are becoming colder. In bed, I've taken to wearing a balaclava over my sparsely covered pate. As well as keeping my head warm, the balaclava stops my ear plugs falling out, and adjusted to cover my eyes, it convinces the restless brain that it is actually night. I've even tried the balaclava pulled right down and reversed, bank-robber style, but found that method somewhat claustrophobic. These night-time strategies would be very effective if everything stayed in place but inevitably, as I toss, my nightcap comes adrift. I wake, concerned that the thrashing tent is being shaken from its anchors, only to realise that the balaclava is somewhere under my stretcher and one ear plug has somehow migrated to the bottom of the sleeping bag.

The blizzard has gone. The barometer is up and continues to rise. Rookie confidently predicts a perfect day. The walk to Mawson's Hut is a delight as I shuffle through fine, shin-deep powder snow for most of the way. Everywhere the patterns of the wind are freshly inscribed on its surface. The powder is so dry that tiny snowballs form, dribbling away across the slope as I churn onwards through the drifts.

The hut is a transformed building. A shapely drift tail has been left in its lee and icicles hang from the yet to be trimmed Baltic pine boards on the workshop roof. Snow covers most of the artefact field, and a regular grid of small timber stakes placed to help Estelle's mapping, is all but buried.

Today is sponsors' day. It's my job to photograph corporate banners draped over the Weatherhaven or held aloft by Ted and Dave Gillott on the roof of the hut.

Our wine and beer sponsor, Yalumba, has entrusted us with an oversized wine bottle, that David Jensen referred to as a 'Methuselah'. Sporting the Veuve Clicquot label, it is alas, only a prop but if it were full it would contain the equivalent of eight regular champagne bottles. My instructions are to take some publicity shots in front of the hut or among the penguins. The bottle is almost twice the height of an Adélie. Carefully placed beside a gathering of Adélies on the Boat Harbour ice edge, the bottle proves to be of absolutely no interest to these normally very inquisitive birds.

Like our expedition, the AAE was funded from both private and government sources. The Commonwealth, New South Wales, Victorian and South Australian and British governments all donated funds to the expedition. In total £25,000 was solicited by Mawson from the five governments. A further £8,000 was voted by the Commonwealth for the second voyage of the *Aurora* in 1913. A plethora of private donors provided both cash and kind. Many of the major donors were leading Australians of the time: Dame Nellie Melba, Samuel Hordern and Hugh Dixson among them. A number pledged donations of £1,000 and more. As explorers are want to do, Mawson acknowledged his benefactors by assigning their names to newly discovered geographical features. Hence, Cape Denison honoured Hugh Denison, a prominent Sydney businessman and the Mackellar Islets commemorate a British patron and donor of the expedition's library. The name Commonwealth Bay, of course, records the support of the governments of the newly federated nation.

Some of Mawson's sponsors are also assisting our project. From Jaeger, a British manufacturer who provided woollen clothing to the AAE, we've each received a down-filled jacket that wouldn't be out of place on the most fashionable of ski-slopes. Drugs and medical equipment were supplied to the AAE by Burroughs Wellcome. Their descendant company, Glaxo Wellcome have maintained the tradition and given this expedition specially inscribed personal first aid kits. And who could forget Yalumba, perhaps our most appreciated supporter, and sponsor of Mawson's third Antarctic expedition, the 1929–31 BANZARE voyages.

After a late but tasty lunch of fresh rolls, bacon and cheese, delivered as usual by Dave 'Pie-cart' Killick, several of us spend most of the afternoon digging the Weatherhaven and stockpiles of materials out of the drift.

Eventually, I'm happily diverted with camera in hand to record some external details of the hut. The low light is perfect. Every fissure, knot and splinter is cast into golden relief and as the hut's timbers warm to the afternoon sun, small drifts of snow melt on the roof, colouring the boards in patches like a fresh coat of varnish.

From behind the living section, with the light streaming over Memorial Hill in the west, the hut's cladding almost resembles the stretched and desiccated skin of an animal. Up close, this ravaged skin tells much about a life of abuse and more than a few stories of the hut's occupation. Marking years of wear, wasted nails stand well proud of the timber surface, and where some external makeshift battens and cleats were fixed, only raised 'shadows' remain, the result of differential abrasion of the underlying cladding. Other battens defy the wind and cling to the hut, a number still fringed by telltale fragments of sail cloth.

From the moment of its occupation, the hut's interior was plagued by fine drift, forced in through the tiniest chink or gap in the cladding and ceiling lining. Once inside, it would settle like dust, then melt. It became an enduring annoyance that prompted a variety of remedies. The most effective was the wholesale wrapping—'Christo-like'—of the southern end of the hut using an old sail left by the *Aurora* on the relief voyage of early 1913. The sail, except for a few tiny fragments of fabric still held down by battens, has long since been clawed away by the wind.

I walk back to the Sørensen alone, relishing the solitude of a rapturous evening, while studying the record of life left in the snow since the passing of the blizzard. Just as our comings and goings are stamped in the fresh drift, so too the brush of stiff penguin tails and the mark of small but purposeful, clawed feet.

Yet another night for *al fresco* dining, with Joan serving pork sausages, pasta and vegetables, followed by reconstituted, freeze-dried yoghurt. Despite the fine evening and the prospect of a painting or photographic sortie, my body protests. My last bath was ten days ago, so I relish the luxury of only my third full wash since the *Shokalskiy* left. The air is totally still, so I leave the door of the Apple ajar and let the sun stream over my increasingly pallid body, while enjoying what must be one of the world's best bath-house views. The relief is memorable and for the first time in many nights, I sleep long and soundly.

Tuesday, 13 January

A light overcast and a gentle breeze from the south-east herald a productive day. I haven't done much painting close to the shore as yet, so I decide to head for John o'Groats by the coastal route. It's an easy walk over alternate bands of snow and rock, past a few small colonies of penguins and their attendant skuas. A couple of ridges over from the Sørensen, just below Low Lake, I find protection on the western side of a small bluff. Below, an evenly graded snow chute slopes directly to the shore. With easy access to the water, it's a favoured penguin track to and from the nearby colonies.

I settle to work downwind of the usual astringent mix of ammonia and fishy brine. After getting paper, paints, water and brushes set up, I realise that I've forgotten to replenish my stock of white, so mildly annoyed with myself, I secure everything likely to blow away and head back to the Sørensen.

I find the 'home team' in a happy mood. Joan is busy making bread, singing along to Patsy Kline on the communal ghetto blaster, while Rookie stamps souvenir philatelic cover, after philatelic cover, from neat piles arranged on the table. As well as radio operator, computer expert and meteorologist, he's also our official postmaster. Cancelling thousands of envelopes with a selection of commemorative rubber stamps is a mindless, repetitive job he doesn't relish. But as the covers are another one of David Jensen's fund-raising ideas for the Foundation, there's obviously no say in the matter.

While I'm looking for my white paint, the satellite phone rings. His concentration disturbed, Rookie grumpily moves to answer the call, but on picking up the receiver says in a cheery voice, 'Good morning. Commonwealth Bay, home of the blizzard'. Answering calls, Rookie can be outrageous and wickedly funny. The one caller to have tried contacting me on our intermittently operating link was told that I'd been 'put down' after repeatedly biting the penguins. Fortunately his tone betrays his humorous intent. I hope!

Within 20 minutes I'm back to my painting and two hours later have completed a satisfying study of the shoreline looking west towards Cape Hunter. I'm particularly pleased with the spontaneity of my sky.

I take lunch at the Sørensen—something of a change for me over recent work days. As well as Rookie, the other members of the 'home team'—Joan, Rod and Dave Killick—are there. Our group seems to be subtly but noticeably dividing between those whose days are spent around the Sørensen and those who frequent the worksite. Like the differing emphases of Alan and Geoff on the conservation program and the wider cause, so our group's varying tasks place some members in smaller gatherings with priorities and attitudes slightly at variance to the whole. As Joan, Rod and Rookie have each spent a year or more in the company of at least one other of the 'home team', their familiar dynamic only serves to reinforce any differences. Thankfully though, overall group harmony is not in question, as each of us remains very much committed to the success of the project.

Estelle is also among the Sørensen lunch crowd. I suppose she's now a *de facto* member of the home team, having been pretty much confined to the hut since her accident. She's getting very anxious about the state of her work. Although a few tentative forays to Mawson's Hut have been made, with Rod as caring chaperone, she still lacks confidence in the wind, and, I think, remains in a deal of pain. Add a bad dose of 'cabin fever' and we have one quite unhappy archaeologist in our midst.

A suggestion that she comes on a painting excursion is instantly agreed to, so this time with company, I again head in the direction of John o'Groats. We find shelter among a large pile of boulders only fifteen minutes from the hut, and as usual, the spot has been used by our neighbours intent on new coats of feathers. Estelle draws one of the resident dead penguins, while I continue to grapple with making sense of the pattern of this place—of how the simple elements of ice, bare rock, sea and sky mix and jostle before the eye, presenting expressive opportunities unimagined at first glance. In a productive burst I produce two paintings. The second concentrates on the glories of the afternoon sky. Extravagant swirls of high cirrus top several opposing layers of what Mawson referred to as 'radiant' cloud—parallel bands which, in perspective, appear to originate at one point, somewhere over the horizon. Planes of grey, cream, blue and mauve are banked one on top of another in a sky that would've had Turner or any of the great Romantics in rapture. My memories of Cape Denison would be made meaningless by the loss of the Antarctic sky, at once ominous or glorious but always restless.

After dinner the table is cleared and carefully wiped clean. We then resign ourselves to a group signathon of the first batch of commemorative cards that will be inserted in several thousand of the philatelic covers that Rookie was working on earlier in the day. No food or drink allowed but everyone makes light of the task with much good humour.

Wednesday, 14 January

As I've not been to Mawson's Hut since Monday, I make photography of progress on the hut and documentation of artefacts my priority for the day.

The living section of the hut is now a transformed space. Within our work area, the last of the ice and snow has been cleared and the ceiling platform lies ready for repair. All artefacts found there have been taken out in frozen blocks of ice for thawing and recording. With the skylights now removed for restoration, daylight can stream into the hut for the first time since 1914. Even Mawson's room is lit. Removal of the ice and snow hanging over his space has revealed a void in the platform directly under the southern skylight. The more we know of the building, the more the logic of its design impresses. Placed neatly beside the opening above Mawson's room, a pair of high, lace-up boots with canvas uppers has been found. We'd all like to think of them as Mawson's, but there are no clues to their ownership other than proximity to his room.

Down below, enough ice and snow has been removed from near the stove to allow me to squeeze into the darkroom. I duck under loose sheets of ice-encrusted tar paper drooping from the ceiling and find the tiny room in disarray. The floor is a mess. Half-frozen rags, rusty tins and a bucket, broken wooden boxes and discarded packets from glass photographic plates make it difficult to get right inside. In the far corner, a sink encrusted with spent chemicals is filled with rusty enamel dishes and broken glass beakers. The shelves are more orderly though, and still hold an intriguing array of coloured glass bottles and unopened packets of photographic paper. Sapphire blue bottles of Pyrogallic acid, a common developer, sport ornate labels with German print: 'Chemische Fabrik E. SHERING Berlin'. Above, on a plain white background under the description of the bottles' contents, is stamped a warning in English, 'Not qualified for internal use'. Other bottles bear labels in French, while one or two are still in boxes showing the Burroughs Wellcome brand.

Closed for years, the air inside the darkroom is heavy with a mix of damp rags, acrid chemicals and tar. It seems hardly credible that the young Hurley produced some of his most enduring images in this tiny disorderly space, barely 1.8 metres square. Here, several of Hurley's life-long artistic themes were first made crystalline. The pre-eminence of nature, ennoblement through adventure, and the value of stoicism and courage in the face of adversity, were all fixed in black and white in this darkroom. As a photographer myself, I feel both privileged and humbled by this glimpse of a true master's working space. For me, it's akin to a musician touching the keyboard of Beethoven's piano.

The AAE offered Hurley his first great opportunity and was the start of an enduring connection with Antarctica. Within a year of being in the company of Mawson, Hurley was again lured south. In October 1914, he left Buenos Aires as photographer on Shackleton's eventful *Endurance* Expedition. The events that unfolded, including escape from the *Endurance*, crushed by ice in the Weddell Sea, and Shackleton's rescue dash to South Georgia in an open boat, rank as the greatest ever feats of polar survival. Hurley was there to record much of the drama. And so continued a passionate life of adventure, embellished by high artistic achievement, in which Hurley photographed the front-line action of two world wars, ventured to still mysterious Papua New Guinea, and accompanied Mawson yet again on the two BANZARE voyages.

Why the darkroom should be such a mess is unclear. Hurley only stayed for the first year of the expedition (although he did return to Commonwealth Bay on the *Aurora*'s relief voyage of 1914), and I'm not sure how much use the darkroom received in his absence. One clue could lie in the problem of Sydney Jeffryes. In his ongoing struggle with the mentally ill radio operator, Mawson often tried to persuade a reluctant Jeffryes to wash. Thinking that a little privacy might act as an inducement, Mawson gave over the darkroom to him.

On 18 July 1913 Mawson noted in his diary:

> Jeffryes a trouble. I spoke to him sharply last evening telling him to keep himself and bunk clean. Had the darkroom prepared for a private washing chamber for him. It seems to have sunk in, for today he has washed out some of his clothes.

On the roof of the workshop, Ted, Paul and Dave Gillott have been installing new metal flashings and a valley gutter before going on to clad the final plane of the roof. It's hoped that these inclusions will prevent summer meltwater from penetrating the junction between the two sections of the hut.

After photographing progress on the roof, I return early to the Sørensen with enough time before dinner to walk a few hundred metres along the shoreline and paint, this time looking east to John o'Groats. A light overcast robs the landscape of shadows or vitality. It also mirrors my own state of mind. Feeling tired and 'flat', I produce a very average study.

As the evening progresses, cloud continues to gather. After dinner Geoff and I climb to the highest point on the ridge just above the Sørensen. In common with a number of such elevated spots I've investigated, it resembles some sort of high sacrificial altar. It's littered with penguin bones and is most likely an abandoned skua's eyrie.

In the silent calm, we take a series of panoramic photographs of a sea the colour of molten lead under a now familiar, though still disconcertingly warm, mantle of grey cloud. Geoff returns to the Sørensen after about an hour, but I decide to try and photograph what I couldn't achieve with my earlier painting. Along the shore, the light is, if anything, worse than before.

I abandon any thought of landscape photography and decide to wander, content just to enjoy the quiet, if sombre night. But there's plenty of interest, as I sit for a while in an area frequented by snow petrels and Wilson's storm petrels. In the increasingly gloomy light, spectral shapes flit and dart, flying in regular passes back and forth above the rocks. Although their nests are close by, the 'Willies' never seem to land.

I can hear much persistent chirruping from somewhere within the boulders and am determined to find at least one nest. But tracing the sounds to their source proves harder than it might seem. After fifteen minutes of peering into dozens of cracks and crevices without reward, I pause to rest on a boulder. A sustained and strong note starts up somewhere directly underneath me. It seems I've sat on a nest. At the base of the rock, in a deep crack that wouldn't be more than 30 to 40 millimetres high, I can barely make out an adult and a downy chick. Although I'm pleased with my find, the crevice is not even wide enough to make a photograph worthwhile. According to Mawson, even the 'ingenious and industrious' Hurley had trouble securing a photograph of a Wilson's storm petrel on its nest, so perhaps I shouldn't feel disappointment.

Back at the tents at 1.30, all's calm and everyone else is in bed.

Thursday, 15 January

We awake to light snow. Because of the absence of wind, the Sørensen fills early with breakfast patrons. The decision to get up on such days is not fraught with the agonising tussle between warmth, comfort and duty that occupies the first minutes of the day if there's a blizzard raging.

With six or seven around the table, the casual reading matter—such as it is—proves to be in short supply. There's the copy of the *Hobart Mercury* from 9 December with Ted's happy face featuring in a picture story entitled, 'Mawson fan will help save an icon'. A couple of days ago, with the paper lying open at the said story, Paul was given to remark, 'Ted, I see you're in the paper again!' This wry quip has been taken up by the group as a comment on our collective isolation, so whenever the now tatty *Mercury* is flicked through, the reader will invariably comment on Ted's repeated appearances and growing celebrity status. If we didn't joke about our isolation but thought seriously about it, we might start feeling morbidly insecure.

There's also Paul's motorcycle magazine and several of Rookie's yachting journals lying about, but their readerships are specialist and therefore small. The publication that receives the most attention is the Christmas issue of *The Australian Women's Weekly*. Lent to the women in our group by Margie, the *Weekly* is more often seen in front of a breakfasting male. The stories of the rich, famous, pretentious and not so glamorous are particularly popular and take on a heightened sense of the ridiculous juxtaposed with our group and its surroundings. And as those same breakfasting males have now been away from civilisation and wives, lovers and girlfriends for more than a month, the increasingly dog-eared magazine seems to open spontaneously at pages that advertise women's underwear.

As the morning progresses, the snow becomes heavier, with large wet flakes quietly splodging onto everything before quickly melting. Over at the worksite, Alan and I run some survey levels from the AAE benchmark (just to the east of Mawson's Hut) down to the shore and as close as possible to where the tide gauge was deployed. From the readings gathered, Alan will be able to establish a starting height against which the data from the gauge can be calibrated. As we work, a soft, whispering stillness hangs in the air as the snow falls ever more heavily. Sounds are dampened by the snow and even the distant babble of penguins is faint.

After a lunch of fresh bread rolls, bacon and cheese with lots of laughs in the Weatherhaven, I start the photographic recording of the first of the artefacts from the ceiling platform. Estelle's back at work and feeling a lot better for it. Rod is keeping an eye on her and helping with some of the recording tasks, so she is now less anxious about not completing her program. Setting up my tripod with camera and flash in the tent, Estelle and I process a miscellany of artefacts ranging from pieces of tissue paper and cotton wool, to lumps of modelling plasticine, and shards of glass and fragments of scientific equipment.

There's also a tiny booklet from the ubiquitous Burroughs Wellcome, who seem to have supplied just about everything of a chemical as well as medical nature to the AAE. Roughly the size of a credit card, the booklet originally accompanied a staining kit. With fine coloured illustrations, it details techniques for staining microscope slides and for recognising the resultant indications of various micro-organisms. Rod is enthralled. As part of the AAE scientific program, blood samples were regularly taken by the medical officer, Archie McLean, and bacteriological cultures made from specimens both of human and animal origin. Studies were made of the survival of common microbes in cold conditions and the natural immunity of the men to infection was monitored. Their absolute isolation was of course a perfect safeguard from disease and they remained in better physical health while at Cape Denison than they might otherwise have at home.

To finish the day, Alan and I take a walk to Land's End, stopping first at the Memorial Cross. With snow sticking to the weatherside of the upright and a heavy grey sky draped overhead, the sombre setting fits the symbolism of the site. Snow also clings to the top edge of the replica commemorative plaque, and as it melts, weeps over the inscription, which reads:

ERECTED
TO COMMEMORATE
THE SUPREME SACRIFICE
MADE BY
LIEUT. B.E.S. NINNIS R.F.
AND
DR. X. MERTZ
IN THE
CAUSE OF SCIENCE
A.A.E. 1913

The original plaque was made from a section of the hut's dining table by the AAE's cartographer, Alfred Hodgeman, and erected together with the cross in November 1913. Badly worn, it was repatriated to Australia by Project Blizzard and replaced with a replica by ANARE.

From the cross we walk along the ridge for a while, observing penguin life as we go. Many of the adults are now caring for very large and healthy-looking chicks with a significant number of twins still surviving. For some though, their brief lives are nearly done. Scattered throughout the colony, a few wet and dishevelled individuals stand on their own, shivering in silence, resigned to their fate.

Adjacent to Memorial Hill, we drop down to the snow ramp that runs south between the cluster of colonies and leads to Land's End. I have more luck negotiating this thoroughfare than at my last attempt. The skuas mustn't be around, so we walk up the ramp unmolested. Further up, the deep snow makes the going hard for the penguins and many of them flop into tobogganing mode, churning through the snow with flippers and feet acting like oars. Not far from Land's End, we notice splotches of blood on the snow. The trail leads us off the ramp and onto the rocks to a badly bleeding penguin. With a foot or lower body gashed, it stands mutely on the edge of the colony, its flippers quivering with shock. How it sustained the injury is impossible to say. Perhaps it was in a particularly vicious fight or just managed to evade a leopard seal. Whatever the reason, the prognosis is grim, as such individuals soon fall prey to skuas. If the resident pair have not yet spotted this hapless bird, they soon will.

At Land's End a scattered group of about fifty penguins lies belly down, nested in the snow near the cliff edge. It's hard to tell if these birds are sleeping or just resting, but it's obvious they're enjoying some respite from the hubbub of the colony nearby where life goes on as usual, 24 rowdy hours a day for the duration of the summer.

We walk back to Mawson's Hut through whispering banks of powder snow. It's a delight to shuffle through its fluffy depths, feet finding a firm footing on the névé (compacted snow verging on ice) about a calf's depth beneath the surface. A pleasure now, but when the wind gets up again the drift will be swept into the air, reducing visibility and adding considerably to the difficulty of just existing in this place. And the more it snows, the more the shovels will be busy at the work site.

Back at the Sørensen we find that the latest call for meat has been heeded by Joan or is it that the contents of the outdoor freezer simply have to be consumed before we leave? Whatever the reason for the choice, the roast lamb, savoury rice and vegetables are very acceptable. Dinner is capped off by a Christmas cake, courtesy of Alan's mother and port, courtesy of our sponsors. The meal complete, and the table cleaned, Rookie pulls out yet another batch of 600 or so commemorative covers to sign. The signings always seem to promote the need for a drink directly after—let's say, as a reward.

Friday, 16 January

The wind overnight is quite light, perhaps only 15 to 20 knots and the temperature a mild -1° Celsius. Joan and Estelle have moved house, repitching their tent adjacent to Rookie and Alan's. It's not that their nearest neighbour, Geoff, snored too much, but as with a number of the tents, a meltwater puddle formed in the middle of its floor. In some cases, tent-mates have become unwilling bed-mates as they roll together in the night, their tent floors all hills and valleys due to the continual freeze and thaw. Fortunately, Geoff and I are still on opposite sides, and our tent floor remains relatively even.

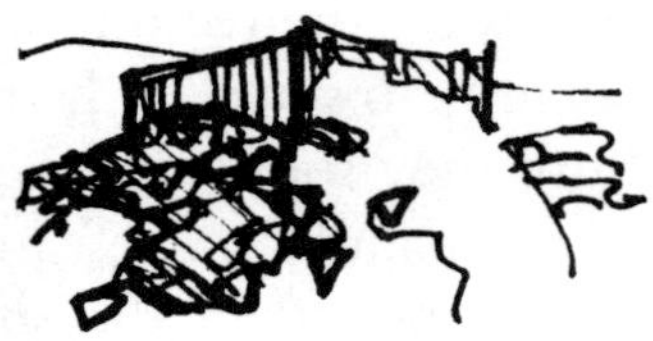

My first stop for the day is the Magnetograph House. Here, I'll do a photographic survey before any repairs are begun. Sitting on flat ground about three hundred metres north-east of the main hut, this tiny recording shelter, measuring less than five metres by two, is in remarkably good condition and the best preserved of all the Cape Denison structures.

Lined with tar paper to make it lightproof, and fortified externally with dry stacked ramparts of rock, the Magnetograph House remains free of snow and ice. It was completed in March 1912, and Mawson records in his diary that the outer walls comprised 'about 45 tons of rock largely pried from frozen positions'. The roof was originally lined externally with sheepskins and hessian bags to exclude all drafts and help maintain an even temperature within. Under the remnants of the pine capping board, small tufts of wool still persist. A snug little building, with a simple lean-to roof facing the winds, the Magnetograph House was built to accommodate magnetician, Eric 'Azi' Webb's most sensitive recording instruments that continually monitored variations in the earth's magnetic field.

The nature of his work was such that Webb had to visit the instruments housed in the two magnetic huts at regular intervals, no matter what the weather, often groping and feeling his way through heavy drift and darkness, there and back. Mawson greatly admired Webb's dedication to the task, noting in *The Home of the Blizzard* that he 'visited the Magnetograph House . . . at least once a day, [and] made his way between various "beacons" by preserving a definite bearing on the wind. His journeys were rendered all the more difficult because they were frequently undertaken at night . . . In darkness there was nothing to recognize'.

Entry to the hut's vestibule is through a stable door. The upper half must have proved very handy when snow was drifted up outside. Just getting into the vestibule is a tight squeeze but the small antechamber allowed just enough room for Webb to slip out of his windproofs and remove his crampons. Because of their ferrous content, the crampons would have distorted Webb's readings, so they were placed outside before he passed through two more outward opening doors and into the darkness.

After managing to edge around the innermost door, I fumble for my head torch. The hut is still as lightproof as when it was last used for science. Although sagging in places and covered with a rime of ice crystals, the lining remains intact. The room smells musty, with a slight hint of bitumen. I pause for a quick look at the hut's contents before taking a series of interior shots with flash.

As could be expected, no scientific equipment remains, but in one corner lies a pile of AAE clothing. Estelle maintains they are 'Mr Webb's underpants'. On her 1985 survey of the artefacts in the hut she deduced that Eric Webb must have been a large man, as the crutch of a pair of long johns she examined, extended to her shoulders and beyond. Scattered around the floor and shelves are various items from French and American visits in the 1950s and '60s: bottles of brandy, ration packs, American soap powder, and, of great interest to Dave Killick, several packets of cigarettes. Both he and Ted have been using our stint away from civilisation and temptation, to quit smoking and Dave has admitted to being sorely tempted by the stale old cigarettes.

Webb didn't remain at Cape Denison for the second year, his recording duties falling principally to Bage. The inner-most of the three baffle doors bears instructions for future parties, left in December 1913 by Bage. Apart from directing prospective occupants to more detailed instructions in a tin on the floor, Bage warns occupants to 'beware of ice from roof falling on instruments', and in a rather ominous turn of phrase at the bottom of the door, wishes them 'good luck and a speedy release' and signs, 'R.B. 18.12.13.'. For the 25-year-old Bage, his speedy release came all too quickly.

A regular army lieutenant on leave from the Royal Australian Engineers for the duration of the expedition, he went on from the perils of Antarctica to those of the battlefield and was killed at Gallipoli in 1915.

I squeeze back out through the narrow vestibule, the doors awkwardly opening on each other, then out into the snowy glare of midday. After photographing each elevation of the hut and its external details, I decide to find shelter from the light but cold breeze that has sprung up in the last hour or so. Drift is also starting to blow. The breeze is uncomfortable, but it's the drift that will be more troublesome when I come to paint.

The Magnetograph House is a short walk from the shore across a flat expanse largely free of ice and snow. Occasional pieces of AAE debris lie scattered across the plain; splinters of timber, the last remnants of rusty tins and although they might be taken as naturally occurring, seal carcasses, their flippers and heads all showing evidence of having been butchered.

It's an easy scramble down to the water, less than a metre and a half below the plain behind. Remnants of the undercut ice foot remain, enough to provide perfect shelter for a couple of hours painting. The littoral seems strangely barren in a region where the marine environment is so rich. Tidal pools elsewhere would be brimming with life. Apart from one sponge-like organism in a single pool, a brief look reveals nothing but a rusty brown smear of algae on the rocks and unruly wreaths of washed-up kelp. Perhaps the paucity of life might be explained by the scouring of the rock platform in the restless growth and decay of the ice edge.

The shelter is perfect, and I'm able to work in a relaxed, gloves-off mood. All I have to worry about is keeping ahead of the incoming tide. It's quite a change from many of my venues here and I think it shows in the relaxed quality of the painting I produce. By two o'clock it's finished, so I head for Mawson's Hut, hoping to time my arrival with a late lunch.

Luncheon fare has been a bit patchy of late. Variety in sandwich fillings has become a concern among the Weatherhaven lunch-crowd and today's offering is typical of the problem. More spam! How the expedition was supplied with a full box of what is ambitiously claimed to be 'luncheon meat' in a tin, is unclear. It's certainly not helping the heartburn sufferers. Also, today's spam is spread with what tastes like jam—though I suspect it's actually chutney. The spam and jam sandwiches are not well received. Mike and Mal respond, filming the ungrateful ones, who peel back the slices of bread to reveal pallid slabs of spam, while grimacing and commenting to the camera—all tongue in cheek, of course.

The day continues to improve, with the wind retreating to a languid calm. At such times, it's a little disconcerting getting used to a changed acoustic: the sound of footsteps in snow; the swish, swish of ventiled legs rubbing as one walks; and even the sound of breathing are all obliterated when the wind roars. The cackle of distant penguins returns and conversations a hundred metres away seem close.

In the western sky, ribbons of light cloud radiate from directly behind the Memorial Cross. In the enormity of the landscape, the cross is minute, yet through its inspired placement, or perhaps even some larger, more puissant force, it seems to haunt the whole of Cape Denison.

The conditions are so good it's time to try painting Weddell seals. Shelter and Weddells don't go together, so the opportunity to work with them is dependent on perfect weather. What a thrill it is to sit out on the permanent sea ice of Boat Harbour and study these creatures. On days such as this, their numbers swell to as many as 50 or so, but when the blizzard rages most quit the ice for the relative sanctuary of the harbour and Commonwealth Bay beyond.

Scattered before me, ten or more seals spill heavy blue shadows across the ice as they slumber on, too sleepy to care much about my presence. Their somnolence is only disturbed by the occasional need for a scratch: deft flipper reaching for basking hide. As the sun tracks toward the plateau, the shadows lengthen and their sleek fur becomes emblazoned in gold and silver. The seals sigh and wheeze, and as if talking in their sleep, utter their strange clicks, squeals and low, guttural rumblings. The lower notes seem to emanate from deep within, making their whole body quiver. Back and forth they talk in a bizarre aquatic language, made all the more strange by me being able to eavesdrop on their conversation.

After a couple of hours sitting on the ice, I've had enough, despite the glorious afternoon. My bottom is numb with cold and my legs quite stiff. I amble back to Mawson's Hut to catch up on progress. The rebuilding of the platform in the living section is going well, with Paul and Geoff working together, replacing the split and broken members. Dave Gillott and Ted are in the last throws of completing the ridge capping and valley gutters on the workshop. From there, they'll move onto the northern plane of the living section roof and make repairs where the hut's chimney once stood. The surrounding cladding was damaged by a fire in 1912 and later repairs have failed, allowing snow to get in.

Ted lifts the repaired northern skylight (temporarily screwed down a few days before by Dave Gillott) and makes a poignant discovery in the oblique cast of the bright afternoon sun. On the top edge of the framed opening over which the skylight sits, he finds an inscription punched into the timber, probably with the point of a nail. It reads 'B.E.S. Ninnis 24 Jan 1912'. A small find perhaps, but one that moves Ted and all of us greatly. It heightens the sense of real people at work, 'personalising' their home. Ninnis may have had a hand in framing the opening or the skylight, and sought to mark pride in his workmanship. Or maybe he was leaving his little piece of graffiti in a slightly furtive way, hidden from view, in case Mawson or the others might not entirely approve. In any case, it's easy to imagine the young and eager 'Cherub' (as he was known) at work on the building. As it was Ninnis's inscription, the most keenly felt impression is one of pathos, knowing that by year's end the young man with the nail was dead.

The walk home to the Sørensen is once more a pleasure, with the soft drift persisting. It makes a change to be able to shuffle through the snow without having to be wary of stretches of treacherous blue ice. For dinner, Joan turns freeze-dried beef into delicious pies with the addition of a light and flaky pastry.

The calm lingers into the evening, so I decide to work on, this time armed with cameras. As January progresses, the colours of the evening strengthen. Swaths of pink and mauve streak the night sky overhead, sweeping in unbroken lines towards their extinguishment on the horizon far behind the the shifting parade of icebergs. As if powered by their own source of light, the bergs radiate shimmering colour long after twilight has settled on the sea. Gold fades to orange, orange to pink and then mauve. I climb to the highest point on a ridge about half a kilometre west of the Sørensen, hoping to capture something of the evening panorama. But I'm too late for the most intense of the colours, so am content to just breathe in the clarity and peace of evening.

I return to camp about midnight, my arrival coinciding with the onset of the evening gale. In the hut, the usual stayers—Alan, Ted and Paul—are still enjoying a drink or two. Quite chilled, I'm tempted to join them. But rather than warm up, then fight the cold afresh, I dump my gear, wish them goodnight and head for the tent. A bitter breeze churns down the valley, driving veils of drift across the fiery sky beyond the plateau. The frigid and the infernal: Antarctica, such a place of illusion.

Saturday, 17 January

It's blowing 45 knots early, with waist-high whirling drift, so it looks like it could be a day to work in my little-used studio.

First though, there's a domestic duty to perform. According to the roster pinned up in the Sørensen, it's Joan's day for emptying the piss and kitchen slops drums into the sea. Lucky Joan! When bad weather coincides with Joan's rostered turn at hard outdoor tasks such as the drum run or water collection, one or more of the chivalrous males in the group will usually double up and relieve her of the pleasure. She's working long hours for all of us, so it's the least we can do.

Anyway, Geoff and I step into the breach on this wild and blowy morning. First, we split the overflowing drum from the Apple with the half-full tin from the kitchen. Stumbling off the Apple's verandah, Geoff nearly loses his load, so we're not off to a confident start.

In strong winds, the trick is to try and avoid the obnoxious contents of the drum being blown over one's ventiles. If the tin from the Apple hasn't been emptied for a while and is consequently near to full, we usually cover it with a plastic bag. On the way to the ice edge, the bag is invariably sucked down into the tin before being caught like a woman's skirt in the wind. Once it's ridden up and off the drum, it's free to flap about, curling around one's arm while flicking last night's urine over the unfortunate carrier's legs. And that's exactly what happens on this occasion.

Complicating our unpleasant duty, Mike and Mal come along to film the epic. Naturally, they get us to repeat sections of the struggle down and back—all for a potential ten seconds or so of film drama. It's with some relief that I finally close the door of the Sørensen, where an appreciative Joan calls me a 'star'. Such are the trials and rewards of the fundamental Antarctic life. But where's Geoff? Mike and Mal still have him captive out in the gale, just near the verandah. It appears they're interviewing him about what it's like doing the bucket run in such appalling weather. They also decide to do an impromptu 'promo' piece for the documentary and have Geoff face to camera, welcoming television viewers to Antarctica. Several of us watch in amusement through the kitchen window as they make take after take of Geoff saying, 'Hi, I'm Geoff Ashley, welcome to Antarctica'—all with correspondingly expansive gestures. The more takes they make, the more absurd it appears. There's also much puzzlement among those still in tents who hear above the wind, 'Hi, I'm Geoff Ashley . . . ' repeated over and over again, like some fragment from an indecipherable dream.

Even though the wind abates in the early afternoon, I remain working inside for most of the day. With much coming and going from everybody and many distractions, it proves to be something of an unproductive and wasted time. I think I much prefer the solitude of my *en plein air* studio, despite its discomforts.

After a day confined to the Sørensen, I head off for some exercise about 10.30 and walk alone to John o'Groats with cameras and tripod. The wind is up again and the temperature must be well below zero. I've noticed the evening cold in the last few days to be noticeably more intense. The summer is all but spent, despite January still having two weeks to run. I struggle to take some shots of the ice cliffs and penguins backlit by a sky flushed pink. Several times the wind almost claws the tripod and camera from my grasp and my bare fingers are quickly robbed of all feeling.

In retreat, I pause briefly to photograph the extraordinary patterns in one of the basalt dykes. Shattered fragments of black crystalline rock protrude from the snow, all aligned towards the sea. Engrossed by my subject, I'm unaware of a nearby pair of skuas. In succession each of the birds swoops, nearly throwing me off balance. But rather than being sensible by submitting to their intimidation and moving out of their territory, I decide to stand my ground and photograph them in flight. Had I a hard hat or similar head protection I might have felt more confident, for after several more close encounters I give up, my nerve severely tested and only one shot taken. Again I retreat and head directly to camp, beaten by wind, cold and wildlife.

Sunday, 18 January

Yet another blustery, noisy night. The gale, our nightmarish lullaby, has become such an untiring opponent in the battle for sleep, I don't know why I keep mentioning it. Perhaps it's just to emphasise the sheer monotony of lying in bed, listening to the tent shake. Ear plugs don't work very well, and in the cold, the batteries in my tape player keep draining fast, so music is no antidote either. Mozart at half speed is about as bad as listening to the wind itself. Whinge, whinge, whinge!!

As it's Sunday, Geoff and I stay in our sleeping bags for as long as we can, not venturing out of the tent until about 9.30. Then, to everyone's delight, the weather rapidly improves near midday. As Rookie says in his situation report e-mailed to Casey, 'Blowing 40kts this morning, a pleasant change I suppose. [?] Quickly dropping out to the most glorious day. Very calm with 8/8 of blue [sky]. Being Sunday t'was a day off for the troops. God exists!!!'

Taking advantage of the calm, I boil some water and wash my underwear and socks, then 'peg' them out on the rocks below the Apple. Soon after my long johns are spread-eagled over the boulders, we hear the roar of a jet engine approaching from the north. Within a minute or so, a Qantas jumbo passes in a broad circle over Cape Denison, then circles again. We were expecting them. Joan does a five-minute radio chat (broadcast to the passengers) with the flight's guest celebrity, Diana Patterson, the first woman to have been a station leader with ANARE on the Antarctic continent. The jumbo then heads south-east towards the Ross Sea. By tonight, when we're struggling into our tents, the Qantas passengers will be climbing into their cars or taxis at Sydney airport after their day trip to the Antarctic. Another over-flight is expected in two weeks. For us on the ground, such events highlight the enormity of change in travel and communications, from Mawson's time to now. I hope they didn't see my washing decorating Cape Denison!

After a quick lunch of biscuits and spreads, Estelle and I think that Rookie's God might be found at John o'Groats. She's still suffering with her shoulder and is finding work painful and frustratingly slow, so relishes the chance to get away from Mawson's

Hut and anything that vaguely resembles an artefact. With one wing clipped she's hesitant about her balance. She finds hopping over boulders difficult, so we decide to skirt the moraines and ridges, walking instead in a wide arc over the permanent ice higher up.

I settle to paint about mid-afternoon, just below my memorable New Year's Day painting spot. The air is still, so I'm able to paint without shelter. Estelle's content to sketch, and we quickly slip into easy conversation about the joys of our surroundings, the progress of work on the huts, our homes, families, friends and acquaintances near and far.

A house-sized slab of ice lies in the water at the foot of the cliffs, having recently crashed from the walls above. On still days such as this, it's quite common to hear the distant boom of falling ice. Mawson comments on it frequently in his diaries, likening the sound to the roar of distant artillery. While we work, a small slab of ice falls from the roof of a cavern not 200 metres across the bay and within a few minutes, a sickle-curved sweep of brash ice starts to drift slowly seawards.

The painting goes well, until a light but frigid, easterly breeze starts to blow directly off the ice walls across the bay. I change from sun hat and shirt-sleeves to windproof jacket, balaclava and numb fingers—all in the space of a few minutes. Fortunately the breeze eases slightly as the afternoon shadows lengthen and I'm able to struggle on with my work for a couple of hours.

Alan joins us for a while and has also come armed with a sketch book. Despite an open invitation to everyone, Alan and Estelle have been the only takers for my little landscape group. Although I'm quite happy to roam the cape alone, accepting my subjects wherever they may confront me, company is always welcome and to be honest, quite reassuring. While I don't take any unnecessary risks and always let one of the others know where I'm going, there is always that thought in the back of my mind—'what if . . . '. Cape Denison is an unpredictable and potentially dangerous place.

Cold and tired, Estelle decides to retreat to the Sørensen in the late afternoon. Alan returns with her, while I remain at John o'Groats to observe and photograph penguins. The Adélie's behaviour, with all its bizarre ritual, coded signals and warnings, is endlessly fascinating. I could easily devote a whole day at a time to just watching their communal life. Try as one might, it's almost impossible not to make comparisons with individual traits in human society; aggression, passion, fidelity, covetousness and so on are all present among the Adélies. And of course, the anthropomorphic image of the little man in formal attire doesn't help one's objectivity either.

There is always a fight somewhere in the colony, with the sounds of squabbling birds rising above the drone of the rest. Even at this late stage in the breeding cycle, penguins continue to fight over nesting materials. Nests are quite rudimentary—no more than a few pieces of gravel or one or two bones of one's ancestors, scratched into a shallow ring. But the raw materials are highly prized. Birds will often steal stones from their immediate neighbours' nests, only to have them stolen back again, while some will even go further afield, embarking on deliberate nest-raiding excursions that often end in vicious squabbles.

The chicks have been growing larger and more demanding of their parents as each day passes. They've begun to gather in small crèches, each crèche under the care of one or two adults. Some chicks are treated with total indifference. One individual I notice, is walked over by an adult, then savagely pecked and left floundering. I can't determine whether the chick is its own, one of its crèche members, or just some pathetic stray. The hapless, bewildered chick has trouble standing, its floppy wings no help, as it struggles to drag itself from the surrounding filth. Exhausted, it finally lies motionless in the mud, like a toy discarded by a child.

I wait near the colony, telephoto lens at the ready, hoping to get some shots of adults feeding their chicks. Despite their rapid growth, it's puzzling that in all the times I've spent observing the Cape Denison Adélies, I haven't seen much feeding activity. Perhaps I've been around at the wrong time of day.

At this stage in the season, both parents are brooding their offspring in short sessions of a day or two. They take turns to feed, swimming up to about fifty kilometres from the colony in search of fish and small marine crustaceans. In parts of Antarctica where Adélies have been closely studied, adults have been found to be away from land for much longer during the incubation phase. Immediately after egg laying, females go to sea for up to three weeks and travel more than three hundred kilometres, before relieving the incubating male. He then heads off for a shorter period and chicks usually hatch during the male's second shift on the nest, about five weeks after the eggs are laid.

Cape Denison's Adélies have not been closely studied. It's unclear how severe an effect the extreme climate has on breeding success, nor is it known if the lack of persistent sea ice helps early in the season by shortening the distance the birds must travel to feed.

Penguin terraces spread over most of the ice-free slopes at John

o'Groats. Carefully winding my way down to the next level, I spot an adult being pestered by two chicks. The more persistent of the pair is eventually successful in getting the parent to open its bill and disgorge the sticky predigested contents of its crop. With camera steady on the tripod, I fire away.

I return to the Sørensen via our outward route, the crunch of hard snow underfoot lending a satisfying edge to the walk. Above the moraine line, sweeping bands of sastruga appear like weather vanes in the snow, their hard, polished facets thrown into bold relief by the dipping sun.

After dinner, Rod, Geoff, the two Daves and I, take a walk inland, travelling about three kilometres up onto the lower slopes of the plateau. For each of us, it's our first foray southward from Cape Denison. Only at this distance does the immensity of the place start to really become apparent. The Mackellars all but disappear over the brow of ice between us and the coast. Further out, the

nearest icebergs appear finally to be anchored firmly on the sea, rather than seeming to levitate on the horizon. A faint breeze starts to spread its silvery fingers over the sea, while overhead, the softening light of evening tinges the sky with muted pastel colours.

Underfoot, the ice seems steel-hard, burnished to a glassy smoothness by the wind. Each of us wears instep crampons, which, with their multiple dagger-like points, only stab into the top few millimetres of ice but grab sufficiently to engender confidence quickly on this alien and treacherous surface. Every foot fall is met with a reassuring crunch.

We walk on past the second brow in the seemingly ever-rising plateau. Effectively limitless and almost totally featureless, the Antarctic ice cap climbs to the south of us in a steady gradient to reach an average elevation of 2,300 metres across the continent. We are only just standing inside the Antarctic circle at 67° South, so to reach the South Pole we would need to walk a further 2,500 kilometres.

What must it have been like for the AAE's sledging parties heading off into this, the ultimate wilderness? The longest of their journeys penetrated 500 kilometres inland, when Webb, Hurley and Bage manhauled their sledge, in fearful conditions, to the vicinity of the South Magnetic Pole and back, between November 1912 and January 1913. Glimpsing just the step from the threshold of that journey, their perseverance and courage can only be marvelled at.

Geoff and the inexhaustible Dave Gillott persevere for an extra 500 metres, at which point they begin to meet small crevasses. With no ropes and only a couple of ice axes between us, we've reached the limit of our preparedness. Going much further might be stretching our luck. In any case, we've had a tiny but tantalising taste of the great Antarctic plateau and can return to camp, well satisfied with our evening stroll.

Our return is brisk, compared to the plodding pace set on the way up. The last section directly above Alga Lake had concerned me on the climb, and I sensed that I might have trouble on the trip back. From a loping stride over the gentler slopes, I'm reduced to a short shuffling gait, fearful of my legs going from under me as the slope quickens. But thankfully, after only a couple of tense minutes, I'm down below the moraine line, just behind the others. We're all back inside the Sørensen for a nightcap at 11.30.

Monday, 19 January

The wind comes up in the night, blowing hard at 40 knots gusting to 50. As Rookie says in his situation report, 'everything is back to normal'. But it's also colder these late January mornings, with -5.6° Celsius recorded first thing. During the night, I realise that I'd left my prized plastic bottle in the Apple. Come morning, it's a wrestle of will over desperation as I struggle to get dressed, fight with the cords on the tent chute, then stumble up the steps to a battle with the door before blessed relief. Just in time!

After maintenance chores, cleaning cameras, logging films and writing a couple of e-mails, I head for the hut valley about mid-morning. The Sørensen as cosy and pleasant as it can be, seems to create its own inertia. It's easy to find nearly half the morning gone before you know it, work tasks interspersed with talk, coffee, and the general hubbub of half a dozen people pursuing their own priorities. No matter what the weather, it's often a relief, just to get away. Once the wind strikes the face and the chill finds its way down the front of the ventiles, there's an accompanying satisfaction that the real day has begun.

Inside Mawson's Hut, Paul, Ted and Geoff are making excellent progress. Repairs to the ceiling platform are just about done, with the fractured collar ties, joists and beams now spliced and connected with steel plates. So that our repairs remain obvious and not be confused with original AAE timbers, Geoff brands each new member with the date and the letters 'AAP'.

The hut's acetylene lighting plant (located adjacent the north-east corner of the ceiling platform) has been cleaned and reinstated to its own supports. During the AAE's occupation, acetylene gas was produced in the plant by dripping water onto calcium carbide in a similar manner to early miner's lamps. The gas was fed through flexible metal pipes to portable lamps that lit the entire hut. The rate at which the water and carbide mixed had to be carefully watched as overproduction of gas had potentially explosive consequences.

After taking a series of flash photographs of the platform repairs, I pay a visit to the Granholm or 'Dave's joinery shop' as it has lately become. Dave Gillott has stripped and repaired three of the skylights, replacing panes of glass, glazing bars and beads as necessary, breathing new life into weather-ravaged components. In the case of the fourth, he's built an entire new window. It's beautiful work. With a kerosene heater in one corner, the snug hut is a welcome haven, its warm air fragrant with the aroma of freshly worked cedar, like potent incense from a distant, exotic world.

Dave is typically modest about the simple beauty of his repairs. He's more relieved that the supply of glass hasn't run out. At first he tried cutting glass over at the work site, with sheet after sheet breaking erratically due to the cold. In the warmth of the Granholm, the glass breaks along Dave's scored lines, all straight and true as intended. It's hard to get the man to smile for a photograph but after several warm-up shots, I coax a grin from Dave posing with his handiwork.

After lunch in the Weatherhaven with all the crew, I help Alan deal with the consequences of some of our nocturnal activities—box after box of empty beer bottles that he's towed over from the Sørensen. Together we unload the sled, stacking the boxes among the rocks next to the western side of Boat Harbour. From there it'll be a short carry to the waiting boats when we come to load the *Shokalskiy* for our return. While the beer has been a welcome lubricant in the social life of our small community, the empties have been nothing but a curse.

Even now, with many of the bottles securely stacked ready for back-loading, they're still a concern. Joan considers them to be an embarrassment, saying that they despoil the Boat Harbour waterfront. She feels they'll create a bad impression with the tourists when they first step ashore. Alan and I take a more sanguine view, and in any case it shouldn't really be any of their concern. We were provided with the beer, so what were we supposed to do with it? Empty or full, the bottles would have to go back on the ship. We both feel sure that they won't pass judgement, provided the whole unfortunate pile is removed and not a skerrick of glass or cardboard left behind.

The problem is not the quantity of beer we've consumed, but the totally inappropriate packaging. Canned beer would have been lighter to bring ashore, and the empty tins, once crushed, could fit into a couple of plastic bins. Even to get our empties this far, the boxes have required careful handling, each being saved and then back-loaded with empty bottles. Out in the snow, the boxes disintegrate, so they've been accumulating under the dining table or in a stack under the Sørensen. Ted has assumed the role of volunteer bottle-o.

After helping Alan, I head for Anemometer Hill to attempt a painting of the whole of the hut valley, with the buildings described in their landscape setting. I've thought about painting or drawing the huts in detail for some time but have baulked at the prospect, not knowing what my particular insight could bring to them as a subject. It may seem paradoxical for the architect in me to view this as a difficult task. Sure, I could draw quite accurate renderings of the buildings showing every wind-ravaged splinter, but I feel that photographs can more powerfully express something of the details of their abuse by nature.

It's only when the huts are viewed against the portent of ice, sea and sky, does the intelligence of both their design and siting fully reveal itself. Only in the context of the landscape can their very existence be marvelled at. These are the things I hope to convey.

What came to be known as Anemometer Hill is the highest unobstructed point close to the main hut. It was here that the AAE set up their meteorological station. In *The Home of the Blizzard* Mawson described its establishment in February 1912:

> Madigan was to take charge of the meteorological observations and assisted by Ninnis and Mertz he erected the two screens and mounted the instruments. Special care was taken to secure the screens against violent winds. Phosphor bronze wire stays, each with a breaking strain of one ton, were used, attached to billets of wood, driven into fissures in the rock. Strong as these wires were several breakages had to be replaced during the year.

And so Cecil Madigan as chief meteorologist began the burdensome task of collecting daily weather data. Mawson recounts that:

> Every morning, Madigan visited all the meteorological instruments and changed the daily charts; at times having to feel his way from one place to the other. Attending to the exposed instruments in a high wind and low temperature was bad enough but with suffocating drift difficulties were increased tenfold.

In his diary entry for April 8, Mawson writes that 'Madigan again loses records—tin box blown away on one of his falls. Anemometer now so clogged and frozen up that little further use'. The instruments constantly break down and Madigan suffers for his diligence. Mawson notes on 22 May that 'Madigan gets bad frostbite on his face. Before discovering the fact he endeavours to throw away part of his face, believing it to be a piece of ice.'

As I search for the best shelter just in the lee of Anemometer Hill, a few reminders of the AAE weather station lie wedged in the rocks about me. Splintered slats and louvres, and strands of the phosphor bronze wire are about all that remain. In the warmth of the afternoon, nestled among the rocks, it's hard to imagine the fury of the blast, the darkness and the unimaginable cold in which Madigan and his assistants operated each day; of May 1912, for instance, when the wind blew at an *average* for the month of 47 knots and the coldest temperature recorded was -35° Celsius.

By 7 o'clock I'm well satisfied with my painting and pleased also that at least I'm starting to think more of the huts in their landscape setting. Commonwealth Bay at its silvern best and streaming ribbons of high cirrus cloud certainly help.

A dinner of minestrone and corn bread followed by cherry and coconut slices make a good foundation for another evening 'signa-thon'. With card after card to sign, I'm amazed that I can produce so many variations on my signature. We're now half way through—only another 2,500 to go.

Tuesday, 20 January

The wind continues to rage and must have blown at more than 50 knots for most of the night. I had a couple of anxious moments thinking the tent was about to be shaken apart. Vivid dreams seem to accompany nights of broken sleep. Last night's was exceptional, both for the fact of recalling the dream and the circumstances that followed. I dreamt the classic snake dream where a large python wrapped itself around my body—sleeping bag, stretcher and all—and began to squeeze. I woke just as it opened its hideous mouth. I'll let the Freudians deal with that one! What's more unnerving than the dream, is that immediately upon stepping inside the Sørensen, I overhear Alan talking on the phone to his wife, Kerry, about how their dog had just been in a fight with a large snake, one that she thought could be a python.

About mid morning I make the all-too-familiar struggle against wind and cold on the walk to Mawson's Hut. Quite an unpleasant day looks likely. All of us are being worn down by such conditions, as we battle just to get around, let alone perform our various tasks. I'm now tired beyond the remedy of a good night's sleep and I know that's a feeling common to all of us.

Despite the privations and fatigue, I'm determined to keep going to the end, maximising every opportunity. Above all, I'll keep painting, even in the most marginal conditions. It's here that I'm finding great exhilaration and the work I've accumulated so far is possibly some of my best. The more I work out in the cold and the wind, the less concerned I become. If it's possible, I think in some small way I'm becoming inured to the sting of the weather. The reward at the end of each day helps. I'll always remember the walks home to the Sørensen with a new painting in my portfolio, when my step would quicken and I'd be content with the feeling that I'd found one more precious fragment that was going towards the making of my own view of this incomparable landscape.

The work site is scoured by a 40-knot wind and despite the bright sunshine, it looks bleak and deserted. Geoff, Ted and Dave are over at the Magnetograph House, inspecting the building, prior to recladding the roof and making a few minor repairs. I don't think they'll be working there today, it's just too windy. I look inside Mawson's Hut and find Paul refixing some shelves in the living section, brought down when the ceiling platform collapsed. As is usual on my rounds, I take a few shots of work in progress.

In the Weatherhaven, Estelle is trying to make sense of the strange mélange of objects and substances that have been slowly melting from the bottom layers of ice retrieved from the platform. With the wind persisting, I'm content to help Estelle by photographing her latest collection of artefacts. Estelle presents me with a range of items that she readily admits only an archaeologist could be interested in: pieces of moulting sandpaper, fragments of cloth, rusty bolts and screws. We joke about the heritage value of these fine examples of hut detritus. But as completeness is essential to recording for it to have any value, I press on.

For a break, I go in search of a vantage point for a second painting of Mawson's Hut set against the landscape. The western side of the hut valley, between Memorial and Azimuth Hills looks bitter. Wispy tails of drift curl in the lee of boulders, a sure sign that it's bearing the full blast of the south-easterly gale. The only alternative lies directly south. The going is slow and laborious, directly uphill and into the wind to the first prominent pile of rocks, but as it includes boulders as large as the Transit Hut, there's a strong chance of finding shelter. Downwind from the rocks, soft snow has been sculpted into a steep-walled bowl. Shapely tails of drift spill from its edge, finning north with a grace that seems at odds with the brutality of the wind. As I finally slump into the snow and plant my back against the sun-warmed face of an enormous erratic, I'm relieved to at last pause out of the blast. It's amazing what a little shelter will do. As the wind surges overhead, I take time to relax and enjoy the view, cradled by the emergent rocks in a small pocket of total calm. Encouraged to return to this spot armed with my paints, I drift back to the Weatherhaven and lunch.

On such fine but cold and windy days, the Weatherhaven is more than just a shelter from the gale. Like in a greenhouse or perhaps more accurately a 'yellowhouse', radiant energy is trapped by its nylon skin, raising the temperature inside to something approaching comfort. My energies are definitely on the wane. Over lunch, I dose off with a mug of soup in my hand, and Alan, complaining of tiredness, is soon asleep on a mattress wedged among a pile of gear at the end of the tent.

More artefacts are lined up for me after lunch. Apart from the strain on my back from bending over a camera and tripod for hours on end, it's actually quite enjoyable. The artefacts, no matter how humble, mundane or unpleasant seem to take on a new life when arranged on a white background with a scale beside them. They're momentarily transformed from the ordinary into something special and rare, each contributing in its own small way to a picture of daily life in the Antarctica of the heroic era. As a reward for my diligence Estelle will from time to time find me something 'fascinating', such as a another piece of laboratory glass or a particularly unusual assortment of washers.

Outside, several of the group dig out timber buried during the most recent blizzard. If this task is left much longer, the snow will consolidate and the timber will freeze in place. I join them for a while. We excavate up to a metre and a half of snow from the stacks, then bundle the surplus timber onto the sled and drag it behind the quad to the rocks below the Granholm. From there it's over the familiar rocky terrain to a stockpile that Alan has established where the Apple had stood. We repeat the exercise twice more. It's becoming apparent that a lot of labour will be needed in tidying up and getting away from Cape Denison. It may be some years before the stockpile is used and in the meantime the wind will be constantly trying to fill Boat Harbour with timber, so we're particularly careful about securing the stack. The anchor bolts that held the Apple's guy wires secure are again put into service, and the timber stacks firmly tied to the rocks.

About 4.30, I return to the spot I'd reconnoitred a few hours earlier. My timing is perfect. The wind is moderating, and as usual, the waning afternoon grows more entrancing as the shadows lengthen. Jets of cloud stream across a dome-like sky as the landscape is flooded with incalculable variations on blue and mauve. The roofs of the hut, their clean pyramidal lines hunkered down below a protective brow of rock, seem more than ever to be an essential, almost organic part of the landscape. They hold fast to the rock as if spawned by the stone itself.

Settling down to work, I pull my balaclava down over my face and rewind my scarf, firmly tucking the loose ends into the neck of my ventile jacket. The scarf is a favourite, old and slightly threadbare, with ragged ends and a scattering of small holes. It had belonged to my long-dead father, but up until now I haven't had much cause to wear it. Resplendent in the family colours—the McGregor clan tartan—I brought it to Antarctica more for sentimental than practical reasons, but it's proved to be very comfortable and better than any modern synthetic garment issued by the Antarctic Division.

Brush at the ready, I pause to wonder what Dad would have thought of my current exploits. At times of distress or discomfort in my own life, I sometimes think of his sufferings as a prisoner of war in Singapore in the 1940s. Of being hot, dirty and unwashed, sick and hungry for three and a half years. The thought of him and his comrades going voluntarily to war and suffering for an ideal, makes any minor discomforts we may endure seem totally trivial.

My thoughts then return to our final days in Hobart, where, in more than one official farewell speech we were mooted as heroes in the mould of Mawson and members of the AAE. As well-meaning as such sentiments were, I felt slightly embarrassed and perhaps will feel more so if the idea is again given voice when we return to Australia. To think of us as heroes akin to Mawson is to fail to understand the enormity of the challenges that the AAE faced, each day for up to two years. Despite some quite real privations, we've lived in relative comfort for less than two months at Cape Denison. There's quantity, quality and variety in our food, and communication with the outside world is quick and easy. We are physically isolated, but if we were to be confronted by some dire event, assistance, even if not instantaneous, would certainly be forthcoming from the French at Dumont d'Urville.

I return to the Sørensen in the early evening, slipping a couple of times on the fields of blue ice near the eastern arm of Boat Harbour. Patches of blue ice seem to grow and shift each day and snow too is equally dynamic. In the windy and cold, yet sunny afternoon, the snow is noticeably more compact than yesterday. In that short time, névé has started to form in places. But just as the snow and ice changes as this summer wanes, so it can vary greatly from year to year. Estelle recalls that by this time in January on her previous stay, the stretch of glacier between Mawson's Hut and Boat Harbour was an almost continuous sheet of slippery blue ice. Fortunately, it's nowhere near as bad this year.

Don and Margie join us for the evening. They have recently returned from the plateau after abandoning their attempt to reach Madigan's Nunatak. The blizzard of a week or so back dumped a lot of snow and masked dangerous crevasses in their path. They retreated quickly, leaving their tent for later retrieval, about eight kilometres inland.

After a dinner of Scotch fillet and vegetables followed by coffee cake most of us round the day off pleasantly with a few beers. Alan and I engage in a mock argument while trying to devise a new system of weather analysis based on the Beaufort Scale of rating wind strengths and corresponding sea states. Looking at Rookie's Beaufort chart, with calm to hurricane against a scale of 0 to 12, we're a bit disappointed with the description of the wind for the recorded velocity. 'Strong wind', 'gale' and 'storm' have always sounded too bland for Cape Denison, so we devise an intricate scale ranging from a 'zephyr' to a 'vortex' to a 'maelstrom'—simply because we like the sound of the words. Somewhere along the way a 'force 20 zephyr' overlaps a 'force five maelstrom' with the vortex somewhere in between. Alan and I argue at what point on our scale would a penguin blow over, or what force of maelstrom would stop one getting out the tent for a pee. We can't even agree on the conditions prevailing outside. Is that a force 50 zephyr or force 10 maelstrom?

The evenings are getting dim now with a noticeable fading of the light in the hut towards midnight. We resort to candles.

Wednesday, 21 January

Although tired after a night of constant high winds, I start the day reiterating my private resolution not to let myself run out of steam in our few remaining days at Cape Denison.

With portfolio under my arm I leave the Sørensen at about 9.30 and head off into furious gusts for the hut valley. Crossing the ridge near Gadget Hut, the portfolio has a tendency to want to leave for Australia nearly two weeks ahead of schedule. I'm like some ungainly one-winged beast. There's no easy way of carrying the thing. Tucked under my right arm, it acts like a sail, flapping violently and requiring two hands to keep it under control. Under the left, the portfolio is pressed hard against my body, threatening to throw me off balance. Together with my other accoutrements—a backpack and tripod that seem to always be following me—I have the aerodynamic grace of the side of a bus.

Despite the wind, the day is fine and clear with very little drift, so I'm able to finish the archival photography of the three smaller huts. In turn, I photograph the Transit and Absolute Magnetic Huts and take more pictures of the Magnetograph House, in both colour transparency and black and white. Each structure is photographed from set points, one for each façade, with the camera pointing roughly at right angles to the building. Although manipulating the tripod is quite a trial in such conditions, it allows me to maintain a consistency of viewpoint from shot to shot.

The Transit Hut is in many ways my favourite Cape Denison building. A makeshift structure, cobbled together from packing cases and scraps of timber, and originally swathed in hessian, tar paper and canvas, its purpose is still readily apparent from the remnants of its skeletal frame. About thirty metres east of the main hut, it was used for astronomical observations in the determination of Cape Denison's longitude. Here, star sights were taken with a special theodolite known as the '10-inch transit instrument' on loan to the expedition from the Government Astronomer.

The hut was framed in such a way that sights could be made through a narrow slot in the roof and the north and south walls, aligned along the meridional axis. In the second year at Cape Denison, radio contact with the outside world improved and an accurate time signal was obtained, allowing Bage, the AAE's astronomer, to make the first exact determination of longitude in Antarctica. Before the two metre by two metre hut was built, a large wooden stump was stood vertically and a cairn of rocks built around its base. The building was then constructed around it. Anchored in rock, the stump formed a stable, level platform on which delicate scientific and surveying instruments could be mounted. One of its faces bears the badly weathered inscription:

LONGITUDE

9^h [142°] 39' 40" E

Other details are too worn to read.

Only one wall of the hut remains more or less intact, its inside face repaired with sheets of fibro (probably the work of the ANARE party of 1978). The hut long ago lost its roofing fabric and some of the outer Oregon posts are now so weathered that the sap-wood has completely vanished from the timber. In places one can see right through the posts. All that persists is a skeleton of the wood's hard annular rings picked clean by the wind. Pieces of cladding, their ragged edges worn paper thin, thrum in the gale. Honed by the dry air, they play a tense, buzzing tune.

At the end of the usual Weatherhaven lunch break, I am witness to a tense exchange born of anxiety and tiredness between Alan and Geoff. Just as the timbers of the huts are worn and frayed, so now are our nerves. Our departure from Antarctica looms and there's still much to be done. Geoff's priorities lie in completing all our conservation work on the huts and he expresses a great sense of duty toward the heritage community and supporters of the expedition back in Australia. He's concerned that Paul has been diverted too early from work on the historic huts to what he considers to be non-essential tasks at the Sørensen. Tomorrow, Paul will return to the internal fitout of the Sørensen and the commencement of a new toilet and shower at the northern end of the hut. Compounding Geoff's worries, Ted has strained his back and has been ordered by Rod to rest for a couple of days. It's a confrontation that has been brewing for a few days, starting with Geoff's misgiving over the prospect of spending time moving the Granholm out of the hut valley.

Alan, while always being an enthusiastic supporter of the conservation effort, has additional priorities. As the Antarctic Division's official representative, he would earn great kudos by returning to Australia with all major objectives satisfied. To have completed the conservation program, and the establishment of a fully fledged summer station centred on the Sørensen, would be recognised as a major achievement.

Alan's affirmation and reaffirmation of his support for Geoff's program has, in his belief, gone unheeded. In a fit of annoyance and frustration he refuses to reiterate his backing for the conservation effort. Thankfully, after about ten minutes, the argument abates, ending in a cool and somewhat begrudging agreement to tolerate each other's preoccupations. I'm left with a strong feeling that both Alan and Geoff realise that to carry a dispute further could be very disruptive. We now all need to work more co-operatively than ever. Anyway, we're all fully aware that the weather and the weather alone will dictate our agenda for the remainder of our stay.

It's good to escape the tension of the tent and battle the breeze back to the Magnetograph House to finish my photographic survey. Before moving on to the Absolute Magnetic Hut, I take a break and wander down to the shore. In the first visible signs that summer is near its end, a scum of ice has begun to form on the rocks around the high tide mark.

Slotted into a small depression in the rocky ridge behind the Magnetograph House, where reasonably level ground and marginal protection from the wind coincided, the Absolute Hut is today nothing more than a standing ruin.

Like the Transit Hut, it was built from odds and ends and lined with tar paper inside a single layer of pine boards. It therefore lacked the snugness of the well-insulated and double-lined Magnetograph House. Here, Webb and Bage laboured also, enduring long hours in the bitter cold, where bared fingers froze to the instruments in the performance of delicate tasks. Mawson, in describing the magnetician's labours remarked somewhat understatedly that, 'Three hours sitting writing figures in a temperature of forty-seven degrees of frost is no joke'.

The survey of the Absolute Hut complete, I'm left with about two hours before dinner. Tonight, Geoff, Rod and I are dining at the McIntyres'. With the wind now easing, the slopes on the western side of Azimuth Hill look as if they might provide enough shelter. Settling on the broad terraces a few metres above the water, I'm rewarded with one of my happiest interludes at Cape Denison. The afternoon sun warms my face and hands, and I feel a renewed sense of vigour flowing through my body.

I paint quickly, composition, colour and form all flowing easily. The succession of patterns in the landscape—from the muted greys, bleached and weathered ochres, purples and the occasional smudge of rusty brown on the terraces, to the snow-capped islets adjacent Land's End, and the shadowy ice walls beyond—all seem to interlock, in what I think is my best work of the expedition. Penguins splash and play close inshore, and as a small group of Adélies amble past, I receive a cursory critique from one of their number.

The work secure in my portfolio, I head off to Gadget Hut, for the first time feeling real satisfaction, that if I were forced to leave tomorrow, I would go well pleased with my efforts.

Over large serves of spaghetti bolognaise we discuss more dreams and schemes with the ebullient Don and madcap Margie. Watching them together, it's easy to see how they survived a year at Cape Denison with their marriage intact. First and last they are mates.

Our conversation touches on many aspects of living at Cape Denison from the weather (of course) to wildlife. Don's knowledge is extensive. While living here he was able to study and reflect at leisure. Of his many theories, the one I found most intriguing is that of the fate of the AAE's Vickers R.E.P. monoplane.

In keeping with Mawson's embrace of the latest in technology, the AAE brought an aircraft to Antarctica, only two years after Louis Blériot crossed the English Channel. But the plane never flew in Antarctica, having crashed at an air show before embarkation from Australia. It arrived at Cape Denison without wings and was converted into an 'air tractor' for hauling loads up onto the plateau. A makeshift hangar (long ago blown away) was added to the western side of the living section of the main hut to house the machine. Eventually, the motor over-heated and seized while the air tractor was being used to support the western sledging party, some fifteen kilometres from Cape Denison. Somehow, the contraption was retrieved and its metal frame abandoned on the ice just west of the hut. It was still there, melted down into the ice when Mawson paid a return visit with the BANZARE in 1931, but has since disappeared. Don believes that it hasn't blown away and has simply melted further down into the ice and out of sight. He's keen to return one day with a metal detector and test his theory. In the meantime he'll settle for one more foray onto the plateau.

Weather permitting, he plans to collect his tent and abandoned gear tomorrow. He suggests that we might like to join him and, as permission has now been forthcoming from Hobart, retrieve the remains of the dog.

Thursday, 22 January

The temperature got down to -5° Celsius overnight, but the wind dropped off early and a fine, still day looks likely.

Following Don's suggestion made last night over dinner, Rod and I will accompany him up onto the plateau. We'll bring down the dog and Don will reclaim his tent and equipment abandoned after the blizzard. Alan and Mal intend following later in the day on the quad. Those two will jump at any chance when it comes to the more adventurous activities, but they don't need an excuse as the dog's retrieval should make great footage for the documentary. I've dubbed the animal 'Rex', in honour of Rex Moncur, director of the Antarctic Division. Most among us who know Mr Moncur are doubtful as to whether he would appreciate the thought, but it seems to have stuck anyway.

We rendezvous at Long Lake and there strap crampons onto our boots. Don has dragged two sleds behind him from Gadget Hut. He'll set off with the larger fibreglass sled in tow while Rod and I will take turns pulling the smaller, usually hauled by Margie. Both sleds are, of course, empty for the walk up. Don's will be used to bring down his gear, while Margie's sled will transport Rex.

Starting from behind the lake, we've no time for our calf muscles or tendons to stretch. There's only one way, and that's up—steeply and quickly. The plateau rises directly before us, its base gathering force like a great wave. We're soon walking on bare blue ice brushed free of snow by the wind, where, without crampons, standing let alone walking would be impossible. Even towing a light sled, one can only but admire the tenacity and endurance of the AAE members and their dogs, as they hauled load, after heavy load (often weighing about 250 kilograms and more) onto the plateau, usually in far less agreeable conditions than we've been blessed with.

With the expedition pinned down by the relentless winds through late summer and winter, Mawson's ambitious sledging plans were postponed until the spring of 1912. Desperate to make at least some progress, Mawson began a foray south, accompanied by Madigan, Ninnis and a team of dogs, early in August. Reaching the top of the steep gradient leading up from Cape Denison, Mawson decided to establish a depot and shelter to service sledging parties that would radiate outwards to the east, west and south, and to provide a target in the vast nothingness of the plateau for groups returning to camp.

In *The Home of the Blizzard*, Mawson talks of the establishment of this staging post, some eight kilometres from the hut and at an altitude of 600 metres:

> . . . we determined to excavate a cavern in the ice under the surface of the glacier, for such a refuge from the hurricane would be of great service to all future sledging parties. [After a day and a half it was done. Mawson again recalls:] It was a great relief to be in a strong room with solid walls of ice, in place of the cramped tent flapping violently in the wind. Inside the silence was profound, the blizzard was banished. Aladdin's Cave it was henceforth called—a truly magical world of glassy facets and scintillating crystals.

The party pressed southward for a day into appalling winds, before returning to Aladdin's Cave, and were trapped by a blizzard for a further day and a half. On 15 August, Mawson and his companions made a dash for Cape Denison, turning the dogs loose to find their own way home, lest they career out of control down the slope, with the sled in tow. But rather than follow, the dogs vanished and the trio arrived at the hut without them. More bad weather intervened before a dog-relief party could be mustered. Bage, Mertz and Hurley were finally able to reach Aladdin's Cave on 21 August. Mawson's diary entry for 25 August notes that the dogs remained at Aladdin's Cave throughout and:

> made no attempt to break into the provisions, . . . [but] just curled up in the snow and waited . . . Grandmother was in the worst condition and died 4 hours after relief. He had gone without a meal the day before we left. They were all listless on arrival and would not eat until taken into the Cave and given warm hoosh [sledging food]. The party spent a crowded evening with all [six] dogs in Cave . . . The dogs were all within about 1 day of expiring—all weak in hind legs . . . Grandmother was with difficulty forced to take a few mouthfuls of food. Had a fit, pawing the air . . . An hour before death, looked straight ahead with glassy eyes.

Bage, Hurley and Ninnis returned to the hut on 25 August with the rest of the dogs, presumably after burying Grandmother. While other AAE dogs went astray or died close to Cape Denison, (one was thought to have fallen into a crevasse while roaming free) none of the locations or circumstances are close enough to link them with Rex. Therefore, we strongly suspect that we will be retrieving

the remains of Grandmother, aka Rex.

We travel over some soft sastrugi and then another steep pitch before the slope eases. A couple of hours further on, we sight Don's bright red tent, a note of warmth in an otherwise unrelieved wilderness of shimmering blue and white. Closer to the tent, narrow drifts of snow, like 'spackle' on a plastered wall, conceal small cracks in the ice. Don's tent is pitched on the edge of an expanse of mildly contorted ice, quite different in character from that seen lower down. While it's not a full-blown crevasse field, Don suggests being more alert, as there's still a patchy cover of snow from the recent blizzard that could possibly hide more than mere cracks. His camp is roughly at the upper end of the dam created by Cape Denison, where the rocky peninsula retards the northward movement of the ice cap, and so prevents sizeable crevasses from forming.

Once at the tent, I'm eager to press on and find Rex for myself. Don points to three dark specks on the horizon, about 400 metres away, and about 150 metres from one another. Two are AAE fuel containers and the dog lies between. Aladdin's Cave must have been somewhere close at hand, but with movement of the ice in the intervening years, it would be next to impossible to find any other traces of it today. Whether the fuel tins dropped off a sled, or were originally left at the cave, or simply thrown away, no one will ever know.

In great anticipation, I continue on while Rod and Don prepare a hot drink and start to tidy the camp. Mindful of the dangers, however slight, I don't let anticipation get the better of me, and so probe for small crevasses with an ice axe in those places where I can't avoid walking in snow. On more than one occasion, the ice axe sinks into the snow with no resistance. Crevasse!

Stuck to the ice by a few tufts of remaining fur on the back of its head, the dog is a piteous sight, even as a skeleton. In death, it still appears to cower against the blast—resigned to exhaustion and curled in enfeebled defiance to the cold. Remarkably, some of its desiccated viscera remains within the body cavity, while tendons and sinews are permanently taut, and teeth clenched. It strikes me as a subject to which the likes of Drysdale, Nolan or Tucker might respond: the brutality of life and death in an alien landscape. The deserts of Australia and Antarctica are in many ways parallel worlds, exacting a terrible price on those that defy them.

After a photo session, I return to Don's tent and a lunch of macadamia nuts and hot chocolate from Don's left-over supplies. His selections of freeze-dried food—a variety of complete meals in foil packs—are a far cry from the monotonous fare of Mawson's day. The AAE's sledging rations, consisted of plasmon biscuit (wholemeal flour enriched with casein), pemmican (powdered dried beef and beef fat), dried milk, butter, cocoa, chocolate, sugar and tea. About 900 grams (dry weight) of food was consumed by each member of a sledging party in a day. Measured out in exact quantities and packed in specially sewn calico bags, the food was carefully calculated to provide for the enormous physical demands of sledging. Above all, it was quick to prepare. The crumbled plasmon biscuit and pemmican were pre-mixed and when added to hot water provided an instant beef porridge known as 'hoosh'. But as convenient and lightweight as the rations were, the diet was nonetheless one of bare subsistence, deficient in a number of food groups and vitamins. There was little margin for error and more than one sledging party returned to Cape Denison close to starvation.

After lunch, Rod and I help Don dismantle his camp. Alan and Mal finally arrive about 2.30 and we then all head up to Rex. It's impossible to tell how long he's lain at this spot and whether he's been blown down the slope at some time and then by chance been stuck to the ice, but just a few chips with an ice axe and he's free. The occasion is duly recorded by Mal and myself. Gently placed in plastic bubble-wrap and Margie's bivvy bag, he's then zipped into the sled for the return to Boat Harbour.

We walk back via the fuel tins and a lone stick poking out of the ice. The tins are perfectly preserved in the cold, dry air. Alan chips one from the ice, then lashes it to the back of the quad. It's destined to return to Australia with us and will probably end up on display at the Antarctic Division's headquarters. Finally Rod and I set off to walk the dog, while Don toboggans down atop his sled and Mal and Alan—the two larrikins—scoot off on the quad.

On the way down, I act as brake while Rod guides from in front. The sled rattles easily over the ice but constantly tries to veer off onto the steepest course, almost as if Grandmother was wanting to take the quickest and most direct route back to his home. We both need to concentrate, coordinating our effort and constantly restraining the eager sled.

The Mackellars gradually unfold before us on a dead-flat sea, and from the last hump in the ice slope, the whole of Cape Denison can be seen. We take our precious cargo down the comparatively gentle slope directly behind the hut valley and arrive well pleased with our day's work. Estelle is not overjoyed by Rex's arrival though, as she's concerned about the amount of extra work required to document him adequately. More pressing tasks are crowding in, and her arm remains in a sling, so for the moment, Rex can wait. Still cocooned on Margie's sled, he's left by the work tent for the night.

While we've been away dog catching, Estelle, Geoff and Dave Killick have been busy separating seal skins from a stinking, messy and putrefying stack thrown against the eastern wall of the hut. As a frozen mass, the hides have acted as a dam, holding back meltwater above floor level along that side of the living section. Geoff's been concerned that water may have seeped into the hut over the years. Consequently, he and Estelle have decided to bury the skins away from the hut, in a place where any keen archaeologist can easily get at them in the future. Good to have missed out on that one!

Back to the Sørensen for dinner and Joan's cashew pie, then more commemorative covers to sign. The wind gets up about eleven o'clock and, by the time I reach the tent, it's blowing 40 knots again. Good days rarely come in pairs around here.

Friday, 23 January

I wake, thinking the fine and memorable day up on the plateau was a dream. It's blown monotonously at 40 knots and above for the last ten hours, while the temperature dropped to -4° Celsius during the night.

An early morning discussion ranges between Geoff, Estelle, Rod and myself, regarding the future of Mawson's Hut and the rather exaggerated claims that we're hearing from the press back home—I hasten to add, not emanating from the pen of our redoubtable reporter. We read with amusement, stories of the hut being in danger of collapse—not because of the wind as one might expect—but as the *Sydney Morning Herald* headline of 15 January puts it: 'Mawson's Hut rotting away in "warm" Antarctic summers'. The article goes on to state that 'when the temperature rockets up to about zero degrees, ice around the precious huts melts, causing humidity inside the wooden structures to rise'. I'm sure this has happened every summer since the hut was built (the average January maximum in 1912 was +1° Celsius), and although conditions may be slightly milder now than in Mawson's day, they haven't yet changed that radically. Besides, we've seen no evidence of rot or decay in timber in any accessible part of the hut. On the contrary, the building seems almost as sound structurally (excepting those parts that have failed due to ice loading) as when it was built. Measurements made by Geoff and Alan have confirmed that the structure remains more or less level and plumb.

The irony of the hut's longevity is that if it were just about anywhere in the Australian bush, it would have been attacked by dry rot, termites, or both, long ago. When the Antarctic climate becomes so mild that dry rot and other sources of timber decay are able to take hold, I predict that Mawson's Huts may then be viewed as an arcane obsession in the midst of the catastrophic consequences of uncurtailed global warming.

If the *Herald* article points to anything, it's that there are factions at work, vying for a say on the future of Mawson's Huts. As important heritage properties, their management has drawn passionate debate among both heritage and conservation professionals, and the Antarctic lay community. For years, a small but persistent lobby suggested repatriating the main hut to Australia. Others proposed cocooning it within a protective dome.

Even within more orthodox debate, first Project Blizzard, and now our own foundation, have, at times, provoked widely contradictory reactions from sections of the professional conservation community. Once our project is complete, the looming argument will centre on whether to clear the remaining ice and snow from the living section and workshop and then seal the building from inside, thus preventing a build-up from recurring. I would have thought that the answer might be a simple yes, but the opposing forces are already mustering their arguments. But for now, it's blowing hard and snowing outside, and if anything, the temperature is rocketing *away* from zero.

The morning develops into a hut-bound day of reading, writing and drinking coffee. There's a limit though to the sedentary life, so about mid-afternoon I venture out. Somewhere to the north, a large storm has whipped up a sizeable swell that now pounds our normally surf-free coast. The surface of the sea is all foam and ferment. Rafts of brash ice jostle and churn in the surf, mixing with a swirling brew of brown sludge and skeins of kelp torn from the bottom.

For the penguins it's a difficult time. Small groups come and go, marshalling on the shore, impatient to put to sea and feed. They advance, retreat, confer, advance and retreat again, intermittently barking to one another in an agitated, staccato voice. Intimidated by the waves at one launching spot, they finally march off to try somewhere else. For those caught at sea, the battering is fearful. Occasionally a tiny head pops out of a wave only to be swept from view and not seen again. Penguins are tough little animals, but I suspect this storm might be their match. I notice one bedraggled bird that had either been foolhardy and put to sea, or had been caught by the storm and just managed to struggle ashore. It stands shivering on the rocks, its breast completely red with blood after taking a pounding in the surf. As evening comes, the swell grows. Angry waves assault the nearest bergs and begin to wash piles of brash ice up and over the ice fronts and rocky beaches.

After dinner I head out to take photographs of the chaos. Thankfully, the wind eases and the snow stops. But the evening remains dim—the days of perpetual daylight are on the wane and the winter darkness approaches.

I return to camp after an hour or so and enjoy the pleasures of what is my fourth, or is it my fifth, bath? It's a frightful admission, but my body was really starting to itch and smell. Perhaps even more concerning is how inured one becomes to such basic issues. The previous morning, desperate for a wee, I fumbled the bottle and wet some of my bedding (fortunately not my sleeping bag). I remarked later to Geoff that I was now beyond caring about such things.

Joan and Rod returning from an evening ski, suggest that the view from John o'Groats shouldn't be missed, so with little need for convincing, Alan, Estelle and I head off into the gloom. On the way, we encounter many agitated penguins, frustrated at not being able to go to sea. Some seem dejected, while others are more aggressive than usual and for the first time I'm the subject of a totally unprovoked attack. About halfway across we stop to investigate and photograph a previously located cache of about thirty penguin carcasses, left in the rocks by the AAE. Estelle is intrigued. Why they were just dumped in that location is unclear.

Our walk takes us over what is now very familiar ground. However, the sight of waves flinging great lumps of ice over spots where I blithely worked just a few days before is one I will always remember. So too the scene at our destination, where we gape in awe at the absolute savagery of this place. Brutal seas, surging brown and dirty white, pound the distant cliffs to the east of John o'Groats. As each wave strikes, manes of spume are thrown high in the air, thickening the gloom already heavy with saltspray. Great lumps of ice are gouged from the cliffs and flung into the maelstrom by retreating waves only to be instantly picked up by their successors and smashed against the defiant wall.

Closer to us, the waves are shielded by the eastern end of Cape Denison itself and in the bay immediately below our vantage point, a spreading scum of brash ice dampens the swell. Although not as tempestuous as the distant view, it is nonetheless a frightening prospect. Near by, the spot where I had perched on that crystalline afternoon just a few days previous is still familiar, a tell-tale rock streaked with small splodges of paint. Such pleasures seem so remote now.

We return to the Sørensen via the easier snow route above the moraine line and turn in at about 1.30 after a memorable walk. Twenty-four hours is a long time at Cape Denison.

Saturday, 24 January

We are now in the final throes of our work and given 'average' weather should complete all the program with a day or two to spare. Geoff is now a much happier person. With a still, overcast morning in prospect, it seems a good time to finish the photogrammetry survey of Mawson's Hut for John Osborne at the University of Tasmania.

Apart from the fresh look of the workshop roof, new ridge capping and gutters, and other minor repairs, there's not much to distinguish Mawson's Hut from when we landed six weeks ago. A casual observer might look at the building and question whether we'd done much to the building at all. But in a successful conservation program that is as it should be.

It's inside the hut where the changes are now so evident and dramatic. Dave Gillott has now completed repairs to all the skylights and their protective outer covers and refixed them over each of the roof openings. The covers will remain open during work hours until just before we leave. The ceiling platform is also complete, but its floorboards have not been refixed. Instead they've been left to one side to avoid snow and ice again overloading the structure. And with so much snow and ice removed, light is now reaching corners of the hut, dark for more than 85 years. The living section has taken on a roomy feel and if not for the cold would be a delight to work in.

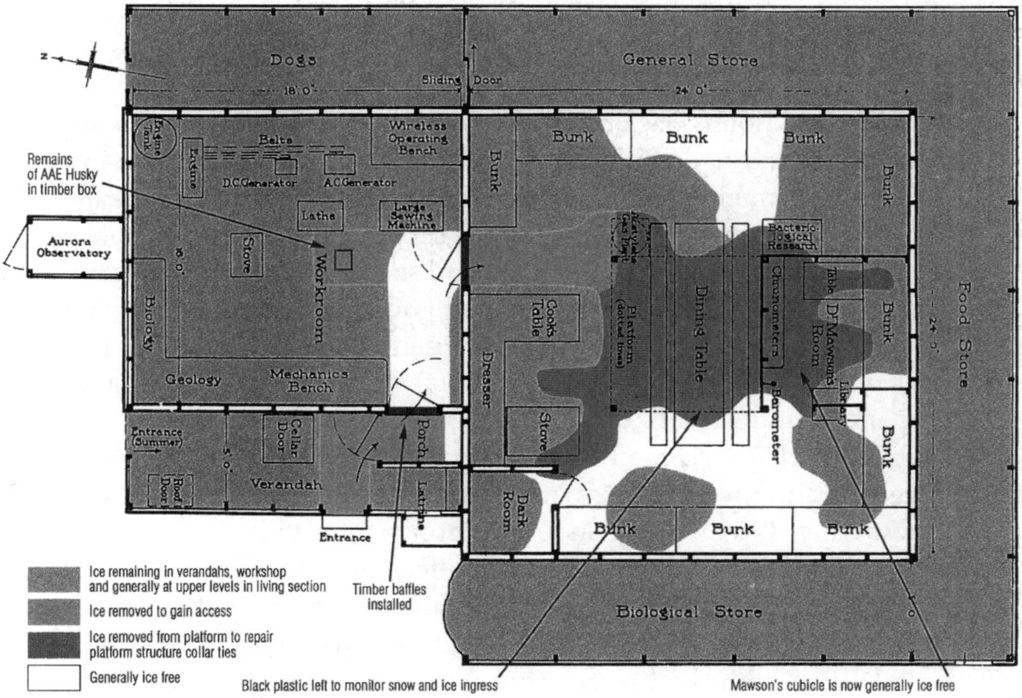

After photographing the south façade of the building, I take a break and head for the Granholm to make myself a cup of coffee. There I find Rookie signing the hut's log book. I also make an entry recording the fact that it's 86 years to the day since 'Cherub' Ninnis inscribed his name with a nail point on the frame of the north-facing skylight of Mawson's Hut. I write that our group have called today 'Ninnis Day' in his honour and that future visitors to Cape Denison should mark it as such. I also note the superb workmanship of Dave Gillott in repairing the four skylights in the hut.

I work into the mid afternoon under dark overcast skies, photographing in overlapping sequence, the remaining façades of the building. It's a slow job—erecting the survey staff beside the building, checking for plumb, steadying the tripod then shooting and moving on to the next camera station—but with the temperature above zero and no wind, all runs smoothly. North-easterlies resulting from a strong high-pressure system offshore have brought relatively warm air to the coast and temporarily dammed the katabatic winds from the south.

After completing the photogrammetry, I had hoped to paint. But as I finish, light and splodgy snow begins to fall, making such a thought quite impractical. Determined not to waste the remainder of the afternoon, I seize on the idea of painting one of my favourite, and as it happens, most sheltered views: looking north from the cold porch of the Granholm.

The Granholm has become a favourite spot for many of us. Small and cosy, it doesn't flap like the Weatherhaven and best of all it has a kitchen, complete with a gas stove. Here, one can escape the cold and the wind, make a toasted sandwich and a cup of coffee or tea. With steaming mug in hand and snack on lap, one can then settle back and watch the comings and goings of the penguins from the luxurious comfort of a folding garden chair. The Granholm is particularly favoured by the 'Brothers' (United Brotherhood of Carpenters and Joiners of Antarctica) aka Dave Gillott, Ted and Paul. This self-proclaimed little group has renamed it the 'Brothers' Lodge' and use it for Lodge 'meetings' (morning or afternoon tea) almost daily. Other members of the group who visit the Brother's Lodge are received under sufferance—feigned, of course.

Initially, I'm pleased with my resourcefulness in thinking to set up in the doorway of the Granholm and the painting goes well for a while. But with the breeze starting to drift again from the north, I find that I've outsmarted myself, and large snowflakes waft in and settle gently on my work. The snow gets heavier, and in the warm conditions, the accumulating drift on the leaky porch roof starts to drip over my painting, materials and me. I struggle on but eventually resign myself to taking home a streaky, wet work.

On the way back to the Sørensen for dinner, I divert for a quick look at the Transit Hut and find that repairs, such as they are, have been completed by Dave Gillott and Ted. Some bracing and wall boarding has been reinstated and a missing roof member replaced. It's hard to know what more could be done to this wind-eroded structure, other than rebuilding most of it. That may need to be contemplated in the future. For now, the prospects for this simple little structure with a stubborn presence are not quite as precariously balanced as before.

It's Saturday night and everyone is in high spirits. We can at last see an end to our work and everyone's sights are increasingly turning for home. The evening gives us a chance to relax, knowing that once we get fully involved with packing up, our opportunities for rest might be limited.

I talk to Paul about his years in Antarctica. He's been coming south since the 1970s and has witnessed great changes in station life in that time. During his first year as a carpenter at Mawson, it was Paul's lot to kill and butcher Weddell seals for dog meat. His descriptions of the job are stomach turning. From that unenviable task, grew an affection for the Weddells and a desire to know more about them. In 20 years, Paul has gone from reluctant butcher to researcher and has in recent times conducted extensive field programs concentrating on the seal's communal behaviour. Despite not having a scientific background, he's become an acknowledged expert on their biology. He hopes to return south at the end of 1998, one of his strong motivations being the continuation of the seal work. In his matter-of-fact way, Paul reconciles past practices with present attitudes as simply 'changing times'.

We continue enjoying a drink or two into the early hours of Sunday. Around one o'clock the 'midnight chompers' come out—snacks of cracker biscuits, pickled onions, smoked oysters, anchovies and cheese. According to those with experience of Antarctic station life, midnight chompers and their accompanying lubricants are a regular ritual. I can believe that!

Sunday, 25 January

Snow continues to fall throughout a mild night. The wind gets up about 5 o'clock but then dies again mid morning. The 'high' must still be around. Most of the group sleep until about ten o'clock, then after a late breakfast-cum-lunch, we all head over to Mawson's Hut for Australia Day photographs. We're holding the session a day ahead of time so that Dave Killick can send a story via satellite, with pictures, for tomorrow morning's papers.

Our high spirits of the past day or so continue. I know we're all feeling quite proud of what we've done and symbolically an Australia Day photo will cap our achievements nicely. The Brothers are now quite a team. As perhaps the most important members of our group, their cooperative spirit and individual resourcefulness has been central to our success. Along the way they've become good mates and have earned the admiration and respect of the group as a whole.

Some time ago, Ted began calling Mawson's Hut, 'Doug's Shack' and with suitable irreverence the name has stuck with all of us. For the main photo the three Brothers climb onto the roof of the shack and appropriately, they are the ones to hold the Australian flag aloft. Don is there as stand-in official photographer, so all our group are able to get in the shots. To add some humour to the occasion, the Brothers do their juggling hammers routine. Each whisks his hammer from its holster, twirls it high in the air, then deftly catches it before it ever looks like crashing through the roof of the hut.

The formalities over, I head for the Granholm and a cup of tea with the Brothers. Several of the support team (those that don't actually stand behind the tools) have been graced by the Brothers with the title of 'Associate'. Of the few privileges flowing to an Associate is the chance to sit in one of the four, highly prized, folding chairs that furnish the Lodge. And what do I do on my first visit to the Lodge as an Associate? Sitting down heavily in one of these flimsy old relics, I break its plastic arm. 'Shame!' cries Brother Ted. Chairs are such a rare and sought-after comfort around Cape Denison, that I've really disgraced myself. The Brothers' punishment is swift: the broken chair is to be reserved for Associates only and the remaining Brothers' chairs are strictly off limits.

My status in the eyes of the Brothers diminished, I walk over to the Absolute Magnetic Hut and do penance by helping Geoff and Alan chip out some of the boards frozen in ice around the base of the hut. They've uncovered the door, which, now only requires levering out of the ice. Like the one on the Magnetograph House, it was originally in two halves, allowing easy entry in heavy drift. It'll soon be rehung on its battered frame. Once enough buried material is retrieved, the hut's structure will be repaired and stabilised in the same manner as the Transit Hut, thus preserving the building as a standing ruin. Alan and Geoff labour on, while I take some action photographs. The snow returns in gentle drifting flakes, swirling around their arms as they work. Cape Denison rolls once more under a heavy canopy of grey. To the west, the ice blink over the cliffs glows with a pallid luminosity, spectacular against the dense knot of cloud out to sea.

Attracted once again by the drama in the Antarctic sky, I head for Penguin Knob and attempt to paint with snow falling around me. Of course it's an impossible task. The wet snow settles on my work and takes all control out of my brush strokes, each gesture bleeding into the one underneath. But I have a solution. With my tripod, I keep a small folding umbrella. I use it for sheltering cameras in these same conditions, so I make a dash for the Sørensen, leaving my backpack to protect a now soggy painting from the snow. I return in time to salvage the work. With the umbrella shaft stuffed down the front of my ventile jacket, I'm able to proceed under shelter with both hands free for my painting. It's a pity no one is about to take my photograph using a brolly in Antarctica!

Returning late for dinner, I find the group in animated discussion over the impending presence of tourists in our domain. The main subject at issue seems to be whether they should be allowed inside Mawson's Hut. Most of us had assumed that seeing the interior of the hut would be an automatic privilege afforded those who were venturing to Antarctica in support of our project. But Joan and Mike don't agree with this assumption.

Joan, in particular, has strong reservations. She feels that it would be too risky both for the hut and its contents, and for the tourists themselves, a significant number of whom are likely to be elderly. Joan doubts whether their experience of Cape Denison would be diminished by a denial of access to the hut. Mike raises concerns about insurance and liability, again with regard to the hut as a piece of Australian and Antarctic heritage, and in relation to the safety of the tourists.

Most of us are quite surprised by their stance. To me it shows a lack of practical appreciation of the consequences. How would we explain to the first boatload ashore that they were free to look at the outside of the building but were not to be allowed inside? Considerable resentment could result on the return voyage, with a real them-and-us atmosphere pervading the ship. It's significant that both Geoff and Estelle, the two people with direct professional responsibility for the building and its contents, are both happy for the hut to receive visitors, provided they are suitably briefed and their numbers controlled.

Joan's approach is just too 'work to rules'. I feel sure that the tourists waived many of their rights to redress when they embarked on the ship. Such is normal when venturing to remote and dangerous places. Also, judging by my past encounters, they'll be a well-behaved bunch and more than deferential to Mawson's Hut.

I can't help feeling that Joan is somehow mistaking Antarctica for the South Australian bureaucracy. Alan eventually diffuses, what is at times a lively discussion, by agreeing to seek Hobart's endorsement—I must confess, at Joan's sensible suggestion.

Monday, 26 January

Blowing hard with drift snow all night and -5.8° Celsius at the first 'obs' this morning. The weather is definitely worsening.

About 10.30 we return to Mawson's Hut for an official Australia Day ceremony. The McIntyres are there too, having brought a flagpole to Cape Denison, especially for the occasion. Don intends recording the event on video for their own sponsors and supporters.

With the help of a couple of our group, the pole is hauled up and securely guyed in place on the large snow bank that continues to grow downwind of the Weatherhaven. Don and Rookie then run a furiously flapping Australian flag up the pole, and into a 30-knot Cape Denison 'breeze'. We had thought of flying the flag from the standard atop the hut roof, just as the AAE had in celebration of the building's completion, but the remaining timber is so worn, it would probably break under the strain.

The group gathers for photographs with Mawson's Hut in the background. As the flag streams in the wind, its edges smacking wildly, 15 hooded figures crowd around the flagpole. Don makes an impromptu speech, saying how much he admires our efforts and recounting how he was greatly moved on first seeing the interior of the hut. The bitter and gloomy morning adds a solemnity to the occasion and a lump starts to swell in my throat. We take turns as photographer, allowing all that want their own record to be in the picture, then after the formalities, celebrate by letting a few snowballs fly.

Rather than starting work immediately, I head over to the Granholm, where I find Dave Gillott and Alan already pottering about. As lunchtime is upon us, I get a chance to redeem my broken chair misdemeanour by making cheese on toast, with coffee or tea for all. Settling to eat in my damaged Associates chair, I feel I could quite happily stay for the afternoon, relaxing and chatting with fellow Associate Alan, while watching Brother Dave work.

With the conservation program all but finished, and no real chance of relocating the Granholm or extending the Apple due to the poor conditions, a few more modest projects have been started. Helped by Alan, Dave is making a couple of plaques to commemorate our work and record the renaming (recently approved by Hobart) of the Sørensen camp as 'Jubilee Base' in honour of ANARE's 50th anniversary. He also intends making frames from scrap timber for some Hurley prints (not originals) given to us by the State Library of New South Wales. The finished items will be hung in the Sørensen, brightening the hut for future occupants.

The warm and welcome refuge of the Brother's Lodge is almost too comforting, but alas, an afternoon of photographic recording inside Mawson's Hut beckons. Despite our group having cleared a significant amount of snow and ice from the living section, much still remains: across the floor, around the cook's table, stove and dresser, and in the north-east corner. Consequently with all that natural refrigerant, working in the hut is really no warmer than working outside. The only advantage is respite from the wind.

In the middle of the space, I set up a makeshift table of a couple of sawhorses, straddled by some offcuts of Baltic pine. I then systematically work my way around the room, taking items from the shelves and photographing them in natural light, as well as with the aid of flash. At least I have company, as Rod and Estelle work away in the darkroom, painstakingly describing, measuring and plotting every item. I'll follow later and photograph the tiny room's contents.

I attend to Mawson's room first. Most of the shelves are still covered in ice, so their contents have to be photographed in situ. But on the open library side, a few ice-free objects remain to be recorded, including a couple of Mawson's sentimental prints—the contentious 'risqué' image, and a moisture-damaged picture of a woman making lace.

Numerous tins of Horlick's Malted Milk are scattered about under bunks and on shelves, their labels proudly announcing that the contents were 'Put up in tin cans expressly for the use of AUSTRALASIAN ANTARCTIC EXPEDITION, 1911–12'. Apart from directions for use, the labels go on to explain its benefits, almost as if Horlick's were a patent medicine. They proclaim it to be:

> Very convenient and sustaining for travellers, especially on forced marches; and excellent for use by Polar Explorers, both on shipboard and while sledging. The Army and Navy of the United States, England and Japan found it invaluable in Typhoid and other fevers, and in all diseases of the stomach and intestine.

Rod is suitably amused by such claims.

By the end of the afternoon's session, I've photographed everything from earthenware jars of nitric acid, to a rusting Bovril Bacon Ration tin and an enamel plate containing what could be someone's unfinished breakfast of congealed porridge.

By the time we head back to the Sørensen, around 7.30, the wind is blowing at least forty knots, so it's with some relief that we burst into a noisy, warm hut, filled with good humour and the smell of a fine Australia Day dinner. Joan turns on some traditional Australian fare: two large sides of beef, baked potatoes, fresh damper and all the trimmings. The McIntyres are in attendance and we all make a great night of it, knowing that this will be our last festivity at Commonwealth Bay. The dinner is also our first big celebration since New Year.

Like us, the AAE relished the chance to celebrate, so any anniversary, however trivial, was remembered. They even commemorated the day when London was first lit by gas. But for the AAE, the occasion of mid-winter's day was of greatest significance. It marked the beginning of the end of the darkness, and the prospect, if not imminent, of spring sledging and eventual relief. In *The Home of the Blizzard*, Mawson recalls the celebrations of Friday, 21 June 1912:

> It was a jovial occasion, and we celebrated it with the uproarious delight of a community of young men unfettered by small convention . . . Hannam and Bickerton shouldered the domestic responsibilities. Their *menu de dîner* to us was a marvel of gorgeous delicacies. After the toasts and speeches came a musical and dramatic programme, punctuated by choice choruses. Outside the wind was not to be outdone; it surpassed itself with an outburst of ninety-five miles per hour.

For such events the interior of the hut was festooned with flags and bunting, and elaborate and decorative menus produced. However, the fare was obviously more fancy in name than reality. For example, the menu for the celebration dinner of 30 January 1912, when the party first occupied the hut, boasted 'Roast Fillet of Dux à l'Adélie [and] Legumes in Season' (roast penguin and tinned vegetables most likely).

We might not have the AAE's creative menus, but I'm sure the hilarity of our celebrations would rival theirs, not to mention the quality of the fare! As befits the occasion, Alan says some fine words about individuals and the group, the collective spirit and our application to the project. Also befitting of such an evening, the celebrations extend a little past midnight.

Tuesday, 27 January

Despite a constant raging gale, I sleep in, having hardly moved all night. I'm getting very tired and at times I think the ship can't come soon enough. Winter must nearly be upon us. I should know, as I stood in for Rookie at the morning weather 'obs' and recorded a temperature of -7° Celsius. With a wind speed of 50 knots, that's an equivalent windchill of -34°!

With the prospect of more recording in the frigid confines of Mawson's Hut, I again postpone work for as long as possible. Alan and I cross to the Granholm in truly awful conditions and join Dave Gillott for a lunch of cheese and smoked oysters on Joan's left-over Australia Day damper. A couple of minutes under the griller—delicious. Two serves of damper and three cups of coffee later, duty finally wins out and I reluctantly leave Alan and Dave to their craft projects. Outdoor work for Alan, Dave and the others looks impossible as the wind now blows harder than ever.

Crossing the open expanse to Mawson's Hut, I divert from my track to take a quick and cautious look at Boat Harbour. Broken ice has been blown into the corner where the Weddell seals haul out and is now starting to set in a surrounding scum of grease ice (newly frozen sea water). At the head of the harbour, the drift spills over the edge like smoke and long icicles grow from the underside of the jutting ice foot. With protection from the wind, grease ice is also forming immediately under the ice foot's protective brow.

Working for hours on end inside Mawson's Hut is proving to be a real test of endurance. Just standing around, my toes quickly lose feeling, so I stamp the snow-strewn floor and fidget constantly in an attempt to stay warm. The back of the camera ices up with frozen drips from my continually running nose and with all this cold and sedentary activity there's a frequent urge for a pee. I knew I shouldn't have had that extra cup of coffee.

So quickly does the urge grip, that at times I've feared the premature onset of some geriatric bladder condition or that my prostate was about to figure prominently in my life. Fortunately, I've been assuaged of any such concerns by learning that the condition is common to all.

One might think the remedy is simple enough—just go outside and have a pee. Not so. For obvious reasons, we've been keen to observe sensible protocols in matters natural around our home and work sites, so the nearest acceptable urinal for us males is the ice edge. On sunny and still days, strolling down to Commonwealth Bay is a pleasant interlude, but as I'm usually gripped with cold-induced urgency, by the time 50 metres of glacier is crossed, a function so simple turns to a nervy race against time.

And then of course, the cold promotes other physiological effects. Exhibiting a measure of justifiable resistance to coming out in the chill air, one's penis shrinks to pathetically unimpressive dimensions. Standing ready, braced against the wind, it's often a battle just locating the thing hidden away in the warmth of several layers of clothing, let alone coaxing it outdoors. Heavily gloved fingers add to the frustration. They're never particularly dexterous. Zippers seem to be their match, and most times, two must be dealt with. Once the ventiles and trousers have been negotiated, a yellow, leather work glove groping about in one's long johns is about as useful and agile as a hand of bananas. The more one tries to rush, the more an impatient bladder just doesn't seem to understand.

Frustrated at the inordinate length of time involved, a bare hand is often the most nimble option, provided one takes care not to drop a glove. Wind and the proximity of the ice edge have nearly meant several losses.

Once the vital organ has been wrestled from its cosy sanctuary one would think that all else was simple. It's then that the gale causes added complications and frustration. 'Pissing into the wind' is a readily understood metaphor for futility, but at Cape Denison, aiming in exactly the opposite direction is just as useless. A bulky, struggling figure standing on the edge of Antarctica is subject to the same aerodynamic effect as a snow petrel's wing. But rather than inducing flight through differential pressure, lift and drag, and all those theories of physics, the urgent pursuit of relief inevitably results in a fine mist, randomly and generously sprayed over one's ventiles.

Mal, as a veteran of such conditions, was quick to proffer advice early in our stay. He suggested the 'three-quarter' orientation. By facing away from the wind a little more than ninety degrees, the flow is quickly whisked into the jet stream and the ventiles saved a dousing.

Finally, having lurched towards the ice edge ahead of the gale, fought with several layers of clothing and assumed the correct orientation to the wind, release from tension is still not at hand. Being blown into Boat Harbour must then be avoided at all cost—the ice edge is a tricky place!

I arrive back at the Sørensen quite late. Over dinner we discuss various options for our withdrawal: when to take the tents down, what projects are still to be completed, will it be possible for some of us to sleep on board ship and make daily forays ashore, while a small crew stays at the Sørensen until the end? Somehow, I feel such discussions can only be academic. The quickening gale outside will determine who goes where and when. The *Shokalskiy*'s schedule allows for her to remain at Commonwealth Bay for up to a week. With very few calm interludes, let alone days lately, we may need all that time to get ourselves and our gear back on board. At 10 o'clock Rookie conducts a sked with the *Shokalskiy*, as he has done for the past few nights. She's making a fast passage and at present is a day and a half ahead of schedule.

Wednesday, 28 January

The morning starts with a moderate breeze and above-zero temperatures for the first time in several days.

I spend most of the morning bent over my tripod in the Weatherhaven, photographing the continuing trail of artefacts from the ceiling platform for Estelle. Many remain congealed in frozen masses. The sheer variety of, dare I say, junk seems to indicate that the platform was something of an out-of-sight, out-of-mind place, where items not in immediate use were thrown in case they ever came in handy—then promptly forgotten. There's more broken glass and fragments of scientific apparatus stuck to old and grimy underwear (I hope it didn't belong to the unwashed Jeffryes), screws, chocolate-box lids, straw packing, unravelling coils of hemp, and tufts of soft dried grass.

The latter was not more packing but a special insulating material known as 'saennagras' that was used to pad the insides of the reindeer-skin boots worn by AAE members when sledging. These fur boots were known as 'finnesko', and, together with their sleeping bags, wolf-skin mitts, ski boots and indeed the Greenland husky dogs and many of the sleds, were acquired for the expedition in Scandinavia.

Although all Antarctic expeditions of the time adopted items of clothing and equipment proven in the Arctic, Mawson, in contrast to Scott and Shackleton, more readily embraced the northern legacy of polar travel as laid down by explorers such as Fridtjof Nansen and the indigenous peoples of the region: the Lapps and Eskimos. Most notably, Mawson persisted with the Scandinavian tradition of exploiting dog power where possible, in preference to what Scott rather vaingloriously regarded as the more ennobling technique of manhauling. It's therefore ironic to note that dogs were almost Mawson's inadvertent undoing, whereas their proper use in the final stages of the march for the South Pole might have been Scott's salvation.

The wind drops enough for me to consider a painting session. Land's End has been a favourite destination these last few weeks, so without hesitation I take the short walk south to a precipitous landscape so utterly different from the broad and gentle slopes of the hut valley.

Skirting the margins of the highest Adélie colony, I'm suddenly accosted by several very aggressive penguins. They come seemingly from nowhere, rushing straight for my shins in a state of great indignity. As I stop, the penguins stand their ground. They roll their eyes, raise the feathers on the backs of their necks and assail me with urgent throaty growls. After a few seconds, I move on carefully and the penguins waddle off in silence, somewhat downcast that all that bravado was to no avail. The aggressive birds seem to be unpaired or without chicks. Perhaps their own brood are now members of a growing crèche, and so without constant responsibilities, they've been co-opted into the colony's own vigilante squad.

Nervously making my way down the near-perpendicular western slope, it appears possible to reach the shore. But the view seems just as promising from a notch in the hillside where a dyke abruptly intrudes, so I settle to work looking back along tumbling slabs of rock to the ice cliffs.

I've now produced 28 paintings of Cape Denison and its environs. Not that quantity is at all relevant or important, but to leave here with 30 works in my portfolio would be very satisfying. Just two more and I can knock off. Light snow begins to fall and the frustration instantly wells up inside me as I sense trouble with number 29. My target is already in trouble. There's no umbrella in my back pocket today and I couldn't be bothered trudging all the way back to the Sørensen to get one. However, my worries are short-lived, for as the clouds drift lower and thicker, sago snow begins to fall. Rather than coming down in wet splotches, the snow falls as hard balls of ice like miniature hailstones. Most bounce off my painting without melting and so my work is again spared any random runny streaks or uncontrolled effects.

With the wind blowing at less than ten knots, I can't help wishing the *Shokalskiy* was already here. Loading the ship would be so easy in such mild conditions, but for now, all one can do is hope that we get a similar break in a few days. There's still a multitude of small tasks to complete if we're to break camp in an orderly way, and perhaps more than ever, the Antarctic factor looms large. The next week could be more difficult than any we've so far faced.

When Davis and the *Aurora* attempted to leave Commonwealth Bay at the end of January 1913 and head west to relieve Wild and his party, they were pinned down by a mighty gale. In *The Home of the Blizzard*, Mawson quotes from Davis's narrative of the events:

> A week's gale in Commonwealth Bay! The seven days which followed I don't think any of us will forget. From the 1st to the 7th of February, it blew a continuous heavy gale, interrupted only when the wind increased to a full hurricane.

On 6 February, Davis noted in his diary that 'the wind became very violent with the most terrific squalls I have ever experienced. Vessel absolutely unmanageable, driving out to sea. I was expecting the masts to go overboard every minute'. Don talks of the whole place 'falling to pieces' in late January and early February. It all sounds so ominous! As in mountaineering, the descent from the summit is often as taxing and just as dangerous as the ascent. Although desperate for a hot shower and a warm bed, I'm not looking forward to the intervening days ahead.

Inevitably the calm is short-lived. A freezing breeze starts to blow off the ice, making the last half hour of painting a real struggle, but I manage to escape with what I think is one of my most confident works so far. One more makes 30!

Back at the Weatherhaven, I warm up by shovelling snow. Mike and Mal are there too and between us we clear deep piles of drift from around the tent. The immensely strong Mal is always active. For him, shovelling snow is as much a convenient form of exercise as it is a necessary chore. Whether this holds true for Mike I'm not so sure, although my enduring memory will be of him repeatedly cutting steps in the increasingly slippery névé that's accumulating between the hut and the Weatherhaven.

About 7 o'clock, the three of us return to the Sørensen and significant news on several fronts. Geoff announces with some satisfaction, that the stabilisation work on the Absolute Magnetic Hut has been finished. Also, the skylights and hatches on the workshop roof have been fixed by Dave and Ted, meaning that the conservation program is at last complete. Only Estelle's archaeological survey is yet to be concluded.

The second piece of momentous news is that less than a day's travel to the north, the *Shokalskiy* is making excellent time through the pack ice. The ship should reach Commonwealth Bay sometime tomorrow afternoon. In an excited, if not slightly apprehensive mood, we settle to what might be our last meal together ashore—another one of Joan's excellent roasts.

Thursday, 29 January

Rodney Russ, the owner of Southern Heritage, absent in December but leading the expedition this time, calls in on the HF radio about 8 o'clock. He reports that the *Shokalskiy* is only twenty nautical miles from Commonwealth Bay and asks that cargo loading start from Boat Harbour immediately after the ship drops anchor. As is typical with katabatic winds, the *Shokalskiy* is only experiencing light breezes. The cold winds draining off the ice quickly run out of puff over comparatively warm seas. Inshore it's the usual story—35 knots, gusting to 40.

Unfortunately, the little bay below the Sørensen is no longer the tranquil cove of mid December. It's too far from the ship in anything but perfect conditions and the shore lacks protection from both wind and wave. All cargo is to be dispatched from Boat Harbour, and although surplus materials, equipment and rubbish have been gradually stockpiled, both there and at the Granholm, much remains to be readied for backloading. The quad and sled should once more come into their own, hauling load after load up to the moraine line, then over and down past Mawson's Hut to Boat Harbour.

The imminent, early arrival of the *Shokalskiy* spurs us into quick action. A breakfast meeting is held to discuss plans, after which we all make a start on packing the contents of the Sørensen. Joan and Rookie attend to their respective areas while the mess room is stripped and all personal gear packed in a flurry of noisy activity. Shelves recently cluttered with reports and files, medical supplies, cameras and the like, soon stand empty, save for the two volumes in the hut's library; *Mawson's Antarctic Diaries* and Don and Margie's account of their year at Cape Denison, *Two Below Zero*. To bolster the choice of reading matter for any stranded itinerant, I leave a copy of my recently published book, *Australia's Wild Islands* and inscribe it with a wish that it might help them remember that warm sands and gentle sea breezes do exist. Borrowing from Bage, I also write that it's my hope they enjoy a 'speedy release'.

With the *Shokalskiy*'s presence just hours away, tensions are rising. Each of us experiences a jumble of emotions as we prepare to be jolted from our home; this savage and frighteningly wonderful place that we both love and fear. Even though we long for the comfort and security of the ship (Mike would walk over water to reach it if he could), the intrusion of others into our world is not an easy prospect to face. Having been bent, humbled and profoundly moved by Antarctica, its presence in our lives will not fade quickly.

For now, there's still work to finish, so about mid morning I leave with Rod and Estelle for more recording in Mawson's Hut. Today, my attentions centre on the long shelves backing onto Mawson's room where much of the Mackellar Library was once housed. All that remains is a single fat novel, a few slim paper-bound volumes, and a pile of magazines and newspapers.

A careful leaf through a stack of twenty or so 'penny dreadfuls' reveals some interesting titles. *To Pleasure Madam* sounds promising, but a brief scan indicates a very tame read. Despite the high humidity inside the hut, many of the magazines appear to be in remarkably good condition. Among the titles are numerous copies of *Punch* and *The Illustrated London News*. Removing a couple for some sample shots, I'm quickly enthralled. I could easily waste half a day just reading. Not that I wouldn't be able to find the same issues in the bowels of any state library at home, but in such a special reading room, I'm easily transported back to the early years of this century.

I chance upon *The Illustrated London News* of 4 May 1912, which is largely devoted to the sinking of the *Titanic*. Generously illustrated, it opens at a double-page spread drawn from information supplied by a survivor, printed in duotone blue and captioned:

> WITH EVERY PORTHOLE AND SALOON BLAZING WITH LIGHT, THE '*TITANIC*' JUST BEFORE A GIANT HAND SEEMED TO PULL HER DOWN. AND SHE SANK IN TWO MILES OF WATER

Like many of the remaining periodicals, the May 1912 issue arrived on the relief voyage of the *Aurora*. What thoughts might have flowed in the minds of those left behind for the second winter as they read of the 'unsinkable' *Titanic*'s demise in a sea brimming with icebergs?

My browsing continues and I chance upon a story, again beautifully but chillingly illustrated, offering opposing medical opinions on the practice of force-feeding women suffragists in British prisons. Perhaps our civilisation has come further in those 80-odd years than we realise.

For a break, I take my usual wind-assisted stroll across to the Granholm, pausing on the way to admire the finishing slices of 'Chainsaw' Ted's latest efforts. To enable us to pass all our cargo down to the waiting boats, Ted has been persevering in the gale, cutting a small quay in the ice foot. He's fashioned a flight of steps down to water level and hammered a couple of mooring bollards into the ice in the form of stout timber pegs. Ted calls this construction his 'stairway to heaven'—the first steps back to the 'babe' he so desperately misses, back to his wife, Cindy. Our ice wharf is now ready to receive the boats, no matter what the tide.

Back inside the hut, I find Estelle and Rod trying to warm up after finishing yet another stint in the darkroom. They've been spending countless hours in its cramped confines, mapping, noting and describing every item. Unable to move sufficiently, they emerge when all gets too unbearable, feet and hands frozen and legs stiff. At such times, the only antidote is to gobble copious quantities of chocolate and jump up and down on the spot—or if one's really desperate, go out and shovel snow.

To end the day, Estelle, with Rod's assistance, moves on to another pressing task. Before Rex is interred, an autopsy must be performed. Estelle's well suited to the job as her acknowledged expertise is in the field of forensic archaeology. In the course of a career marked by unusual challenges, as well as Antarctica and Heard Island, Estelle has also worked in Pompeii, where she conducted the fieldwork for her doctorate on female skeletal remains.

Rex's body is brought into the hut and laid on my photographic table. Remarkably, the remaining viscera in the abdominal cavity is still sufficiently fresh for the room to smell a bit 'doggy' within minutes of his arrival. For an hour or so, Rod and Estelle carefully twist and turn the skeleton, looking for clues as to the life he led, age at death and what killed him. Wasted muscle attachments, a knitted fracture and severely ground teeth, as expected, evince a life of hard toil. And what killed him? Cold and exhaustion, most probably.

The autopsy concluded, I take a series of photographs of Rex before returning him to his very own commodious dog box, recently made by Ted from scraps of Baltic pine. For the next day or so, Rex will stay in the hut. His final resting place is yet to be decided. The Antarctic Division have directed us not to bring him back to Australia, but have advised that he be buried near the hut. Sounds simple. In reality though, it's a question that's been perplexing Estelle and the rest of us for some days. Close to the hut, we risk disturbing artefacts. Further afield, the box may be crushed by moving ice, and his exact whereabouts could prove difficult to track, even with a marker of some kind. I suggested one of the verandahs, where the dogs once lived, but there too Estelle is worried about sullying the archaeological record. She's not keen on putting him inside the hut either, considering it historically 'inappropriate' and too warm for his long-term conservation. Wherever he goes, someone must decide soon.

We emerge from the hut to see an unfamiliar speck on the horizon that can't be confused with an errant iceberg. The *Akademik Shokalskiy* is back in Commonwealth Bay. Although the first sighting of the ship has been a long dreamt of moment, its arrival is depressingly anticlimactic. She is now so tantalisingly close but still out of reach. With the wind gusting at more than 50 knots, this is definitely not an evening for boating.

Back at the Sørensen, Geoff produces the last half of a bottle of Johnny Walker Black Label, that he'd carefully secreted away from temptation in a remote corner of our tent. As most don't seem to be Scotch fanciers, the prospect of a couple of pre-dinner drinks among just three or four of us spurs me into action. Good whisky deserves pure ice, so what better source than the melt lake? Charging off into a 60- to 70-knot blast, I'm quickly doubting my sanity. What's more my doubts are soon confirmed when I slip on the ice and fall with a sickening thud, right kneecap on rock. In an instant, the stainless steel dish I'd brought for my icy prize is making the second of two bounces on its flight back to Australia. After a minute or so, I regather my composure and hobble back to the hut, no damage done except to my dignity. I had feared Joan's displeasure over the loss of what was a piece of her own equipment, but she's more concerned for my knee.

Geoff retrieves the situation by mounting a more successful mission and together we enjoy several hard-won drinks as an accompaniment to freeze-dried something, followed rather miraculously, given the circumstances, by fresh-baked scones and cream. A real morale booster, Joan!

Friday, 30 January

It must be time to go home as my bed has collapsed! The stitching on one side of my stretcher has gradually come undone, meaning that over the past few nights, parts of me have been sleeping directly on the ice. Eighty knots, gusting past 90 and cold—the temperature got down to -8° Celsius during our coldest night yet. This place is turning very nasty. We haven't seen winds like this since Christmas. I remain in bed until about nine o'clock and would have gladly stayed there all day.

Staggering up to the Sørensen, I'm almost bowled over by Ted, rushing off the verandah with a rucksack on his back. I think the prospect of a day barricaded inside the hut with twelve others is too much for him. From inside we watch him struggle up the hill towards Gadget Hut, propping and stumbling as the gusts hurtle down from the plateau.

The wind drops back to about 60 to 70 knots, but by late morning, a blizzard is blowing and there's less than ten metres visibility outside. Most of the day is spent fretfully waiting, catching up on writing and eating badly on leftovers: sledgies, vegemite, jam and cream cheese.

Unanswered questions spin in our heads, occasionally being asked of each other—when will our relief come, and what of the gear we've taken over to Boat Harbour? And the Weatherhaven, our cheery refuge that we've dug out of the snow, time and time again over the past few days, is it once again buried? Was all that effort to no avail?

Everyone is on edge. The easy-going communication that flowed between our group and Southern Heritage Expeditions' staff on the southward voyage has now gone. In radio discussions, Rodney seems tense and anxious. We've heard that serious contractual disagreements have arisen between Southern Heritage and the Foundation, so we're not sure how much of his edgy radio demeanour is due to issues back home and how much is born of genuine apprehension over the difficult task ahead.

During our evening sked with the *Shokalskiy*, Rodney and Alan discuss the provision of wet-weather gear to members of our group involved in back-loading operations. Through some failure in communication, each party thought the other was providing the necessary gear. We'd thought Rodney had ample gear on board and as it wasn't needed when we unloaded, the issue was never in the forefront of anybody's mind. Consequently, no checks were made. When Alan requests that three sets of gear be brought ashore for our boat crew, Rodney is furious. He accuses us of gross incompetence and ends his volley, saying, 'I hope you enjoy your winter'.

None of us are impressed with Rodney's attitude nor his wish for our winter pleasure. It seems that he harbours a rather prejudiced view of our group, yet most of us have barely met him. The black jokes about laying in a stock of penguins had already started without his prompting. The difference is that we're joking our way through a tense time. Rodney is not.

Mike goes into a gloomy depression. Of all our group, he has most keenly felt the isolation. When it looked possible that we'd run short of fuel, Mike saw his beloved e-mail link threatened: how was he going to keep his business going back home? Through careful management the fuel crisis was averted, but I think Mike has felt vulnerable ever since. Rodney's antagonism only serves to heighten his sense of unease.

We again heat up some freeze-drieds for dinner. Ted, who'd not been heard of since breakfast, radios in just as we start to consider organising a search for him. Of course, he'd taken shelter for the day in the Brother's Lodge, and according to his account, enjoyed a meal of freeze-dried steak and vegetables, fruitcake and port.

Late in the evening the wind is still blowing at 60 knots and above. Somewhere out in Commonwealth Bay the *Shokalskiy* is steaming back and forth and there's no immediate prospect of launching the boats. What did Bage say about a 'speedy release'?

As my bed is now useless, I decamp to the Sørensen and spend my first night away from a thrashing tent in just under six weeks. There's no respite from noise though. The wind whirrs through the hut's ventilator, rising in pitch as the gusts mount, then dropping as they ease off. I sleep fitfully, one ear on the wind, and wake early, cursing the leaking air mattress that I'd grabbed from our stores. I'm not having much luck with my bedding of late.

Saturday, 31 January

Sixty knots from the south. It hasn't blown at less than 40 knots for nearly three days now and no one holds much hope of an improvement today. Need I say more. Packing up is continuing as best it can, given the conditions. So with our lives caught in something of a meteorological stalemate, Rod, Estelle and I decide to turn the wait to good use and record as many of the artefacts inside Mawson's Hut as we can. On the walk over, we yearningly scan the horizon, looking for the *Shokalskiy*. She's there, but a long way out, steaming back and forth while the wind remains strong, not willing to chance the uncharted waters closer in.

The photographic task is potentially endless. All manner of fascinating items emerge from every accessible corner of the hut, to be noted by Estelle, then photographed. I record in no particular order: Eric Webb's school or university crest (pinned to the wall beside his bunk), several Leyden jars (large glass cylinders lined with zinc foil, used as electrical condensers), a stuffed or mummified skua, an umbrella, what we think are potatoes, pickle jars, Capstan Mixture tobacco tins, Colonial Sugar Refinery Golden Syrup, a Coleman's Mustard packet, and a lone, broken gramophone record entitled, *The Kaiser Frederick March*. I wonder if the latter's breakage was entirely accidental?

The darkroom is a particularly rich field. Here, there are boxes of French glass plate negatives (some with beautiful multi-coloured labels bearing the brand of the great photographic pioneers Lumière of Lyons), acid and other chemical bottles, and an exquisite match box with a Platypus Brand label. But perhaps the most interesting item is what Estelle believes to be a homemade roulette wheel. Roughly fashioned, it consists of a metal disc about the size of an LP record, with cardboard stuck to each face and a hole at its centre. Both sides are divided into numbered segments.

During the winter of 1912, gambling became a favourite pastime. Chocolate and candles were the currency of the hut and the stakes were often high. 'Bankruptcies' were not uncommon. 'Calcutta sweeps' were run on predicted wind velocities, and as Mawson remarks in *The Home of the Blizzard*, wild speculation was 'redeemed in the ordinary way by the sale of the debtor's effects'. He goes on to explain the background of our find from the darkroom:

> Two financiers [John Hunter, the biologist and Frank 'Hoyle' Hurley], indifferent to the charms of chocolate, established a corner or 'bank' in the commodity. The 'bank' by barter or usurious methods, amassed a great heap of well-thumbed squares, and, when accused of rapacity, invented a scheme for the common good known as 'Huntoylette'. This was a game of chance similar to roulette, and for a while it completely gulled the trusting public.

In the light of all this nefarious activity, it's interesting to speculate on the origin of Hurley's nickname, 'Hoyle'. Perhaps, as in the expression, 'according to Hoyle', Hurley was the AAE's acknowledged authority on anything to do with games of chance.

What can't be easily moved from the darkroom, I photograph in situ. Alas, there are no forgotten gems, such as cameras, lenses or exposed plates left behind in the rush to vacate the hut. Nonetheless, Hurley's presence is palpable everywhere I look. One can almost see his images of Antarctic life and landscape, pulled dripping wet from the now rusty developing tray sitting near the sink—powerful shades of black and white metamorphosed from shadowy negatives. Also, it's the little personal touches that help render Hurley and the others real. Above his tiny, inclined work table, we found the words 'NEAR ENOUGH IS NOT GOOD ENOUGH' written on the wall in pencil. Was this Hurley's own motto or a gentle admonition from Mawson?

While the recording continues, Ted readies Rex for his interment. Rex's precious body is removed from its temporary resting place in the hut and taken outside. There, he's swaddled in still more plastic bubble-wrap, reverently placed in his pine box, and the lid screwed down. Then, 'Chainsaw' Ted gets back into business, and with his usual precision, excavates Rex's 'crypt' in the solid wall of ice that remains in the workshop. Once more in the hut, Rex and his box fit well, sliding easily into the neat, made-to-measure chamber. The hole is closed over with some of the excavated blocks, and Grandmother, Rex, call him what you like, is finally at rest. No more hurricane-force winds for this special dog. Estelle is still not sure that the hut is the correct repository for such a problem animal and I suspect her archaeological sensibilities remain slightly offended. But for now, I think she agrees that Rex can become someone else's problem.

I finish photographing the interior of Mawson's Hut with some overall shots of the light-filled space. With Rod and Estelle working in the Weatherhaven for a while, I'm left in solitude to contemplate my special surroundings for possibly the last time. Somewhere out on the roof, the wind thrums a weary tune on a loose board or a piece of wire. Even within the confines of the hut, the drone of the wind is ever present, forcing its way through gaps in my thoughts as I clamber over pinnacles of ice to reach the far corner of the room. With my tripod set up near Cecil Madigan's bunk, I try to imagine the feeling of dread that must have gripped this dedicated meteorologist as he lay in bed listening to the shriek of the wind, contemplating the chaos that awaited him on his imminent foray outside. Looking around the room to each man's initials inscribed on their bunk, I could easily drift into contemplation of their no doubt similar fears, but also their dreams and aspirations, their longing for home and safety, and how much each sensed of their place in history.

Weary and cold, we retreat about seven o'clock after a solid six-hour stint inside the hut. The mood back at the Sørensen is as tense as ever. Packing has been proceeding all day but still more needs to be done. Despite the seemingly interminable wait, the group's sense of humour persists. I think we all realise that we might not survive without it. We've even ritualised the tedium.

Little difficulties that have persisted as mere annoyances, take on a new meaning. For instance, a kerosene heater usually stands just near the breakfast bar. It's not an ideal place, but there's really nowhere else it can go. Invariably, a newly arrived member of the group while removing hat, gloves, ventile top and the like, bumps the heater. At the slightest touch, this very safe appliance shuts down. So sensitive is its safety switch, that Mike needs only look at the heater for it to turn off. Very reassuring, you might think. But the heater doesn't shut down quietly, it emits a sharp 'thwack', like a rat trap, as the fuel supply is cut. The weary bystander is often startled, and, at a time when our nerves are already raw, the innocent looking appliance provokes a swift retort—usually with the ubiquitous expletive, 'fuck!!!'

This heater kicking routine occurs so often, that the words 'kick . . . ⋆ !!!' have become woven into the black humour of our predicament. While we wait for salvation, the slightest lull in the wind promotes a parroted response from one of the hopefuls in our group, along the lines of 'I think the wind's dropping'. This is followed by someone else saying in disagreement 'No, give it another hour' and after a short pause, a third party will moan, 'I want to go home, nowwww!!!' Straight after one such interchange, Mike offended the heater yet again and uttered the standard expletive. The hut immediately erupted with riotous laughter. Inevitably, 'I want to go home, nowww!!!' has become a prompt for someone to utter in a loud voice, 'kick . . . ⋆ !!!'.

Consumed by anxiety, it's easy to ignore our surroundings, but as January draws to a close, the savage beauty of the Antarctic evening reaches new heights. Colours strengthen as if mounting a defiant stand against the impending darkness. Flushed pink, a full moon hangs in a darkening sky shot with crimson and the ice cliffs are lost in swirls of lurid blue drift.

The long night is coming and none of us want to be here to welcome it home.

Sunday to Tuesday, 1 February to 3 February

I wake in darkness. Rather than a flapping tent or the incessant boom of the wind surging through a vent in the Sørensen's roof, I'm disturbed by the mechanical hissing of air-conditioning ducts. Coming to groggily, my aching arms, legs and back immediately complain and I realise that I've forsaken the floor of the hut for a bunk on the *Akademik Shokalskiy*. How long have I slept? As I grope my way out of bed I'm not even sure whether it's day or night. In the darkness, I stumble over an untidy mess of filthy clothes spilling from my kit bag and lurch for the cabin door, two unfamiliarly bare feet uncertain at the touch of carpet. Desperately thirsty, I then head for the ship's library and a chance to make myself a cup of tea. The ship seems deserted, so it must be early morning. Tuesday, I think?

For the first time since we arrived back on board, I can sit quietly and reflect. We're now heading 150 kilometres east towards the Mertz and Ninnis Glaciers, hoping to cruise close to the tongues of these great rivers of ice. From the library's porthole I can see a slash of orange in the southern sky, a mocking warmth heralding another bitter day as the ship pitches onwards into a stiff easterly gale. If the winds abate, we'll turn back to Commonwealth Bay, after we've seen the glaciers, for one final try at retrieving the last of our gear still on shore: the quad, Dave Gillott's precious tool box, some power tools, the tide gauge and other odds and ends. If the conditions ease for long enough, we might even have a chance of pulling the Weatherhaven down or at least cutting it free.

The last couple of days were exhausting. The wind didn't drop below 40 knots until Monday afternoon when at last we were blessed with eight hours of 20- to 25-knot breezes. In those final few hours, we worked harder than ever. The last of the gear was rounded up from all corners of the site and almost all of it successfully ferried out to the ship. Also, any surplus building materials still not secured were added to the stockpile near the Granholm and the huts were prepared for winter.

It fell to Rookie and me to crew in the boats, with a third crewman, Phil, coming from Rodney's staff. The issue of wet-weather gear was easily solved, making Rodney's radio histrionics seem a bit silly. I had a waterproof jacket and pants with me, courtesy of my own sponsors, Paddy Pallin, while Rookie borrowed a set from Don. Rookie teamed up with Belinda again, while Don and Phil manned the second boat, and Rodney ended up with me as his crewman. Rodney's demeanour hadn't changed since the tense radio skeds of a few days ago. On first landing he was still very antagonistic and launched straight into an argument with Alan, almost as soon as he had stepped ashore. One could only surmise that the dispute with the Foundation had deepened. On last year's voyage both Geoff and I had found Rodney to be quite affable. His recent manner seemed out of character. Whatever the problem, it wasn't going to be solved in Antarctica.

Crewing in breezy conditions was a job I'd not relished. Rodney's driving was skilful and unquestionably safe, but Antarctic boating again proved to be an unpleasant and stressful ordeal. With the wind behind us, outbound trips were always easy. The only problem on those journeys came at the end of the ride with the back-wrenching job of throwing the contents of the boat up onto the ship's cargo deck. Each time we neared the *Shokalskiy*, I would fantasise that this was the last journey, and not only could I see, smell and touch the ship but I would actually be able to clamber on board. Ten minutes later we would emerge from its lee side and plunge into the blast, my idle dreams quickly smashed by steep wind waves and icy spray. With each trip back to Boat Harbour, I was further convinced that sooner or later I'd break a bone or injure my spine.

With the first boats ashore, some new faces finally entered our world. Twenty or so red-jacketted figures stepped onto the Antarctic ice for the first time and our isolation was ended with a few perfunctory handshakes. There was no time for anything more welcoming. The cargo shuttle beckoned. Anyway, the tourists had more engaging prospects in mind than talking to a few weather-beaten and smelly locals, for they too had waited a long time for that moment. Regaining their composure after the spine-jarring ride, they quickly sought out Mawson's Hut, keen to see what all the fuss was about. While the rest of our party shuffled cargo, Estelle escorted the tourists in groups of three or four into the hut. Among their number was Bill Hunter, son of AAE biologist John Hunter. For Bill, an elderly doctor from the Hunter Valley (of course) in New South Wales the experience of seeing his father's bunk inscribed with 'J.H. 1912' was apparently worth all the effort of getting to Antarctica and the tense wait in Commonwealth Bay. After their tour of the main hut, the tourists then dispersed and wandered Cape Denison under the direction of Southern Heritage staff for as long as time permitted.

Also ashore with the ship's passengers was Australian Army

Warrant Officer and explosives expert, Chris Roesler. After lengthy but successful discussions between the Antarctic Division, the Army and the Foundation, Chris had been sent to Cape Denison to deal with a small stockpile of explosives left by the AAE in the rocks to the east of Memorial Hill. The aging explosives were considered to be unstable and a potential hazard to unwitting visitors. Chris's solution was to first carefully move them onto the glacier, high above the last boulder pile. There, he wrapped them in sheets of plastic explosive before detonating the lot, in a single, spectacular implosion. When the deed was done, Rodney and I were nearly two kilometres away from the blast, well on our way to the ship. First there was the puff of grey smoke, then the thump of the blast cannoning off the ice and, almost immediately after, the acrid smell of cordite persisting on the breeze nearly all the way to the *Shokalskiy*.

The first cargo loads ran smoothly, and within three hours we had about half the gear on board—everything from surplus food, medical supplies and nearly 10,000 stamped souvenir philatelic covers, to the somewhat delicate problem of dozens of bags of human waste. Thankfully, the bags were still deep frozen. Under the rules of the Antarctic Treaty, sewage cannot be discharged into the sea south of 60 degrees, so this irksome cargo must remain on board for a few days yet. When the time comes, some lucky people will get to empty the contents over the side before burning the bags in the ship's incinerator. Also back on board were the gas bottles, both used and full spares, and the beer bottles.

The latter didn't help Rodney's mood. Riding back to the *Shokalskiy* with a caged pallet full of empty bottles just seemed to confirm his opinion of our group. What was I to say? It was guilt by association. There seemed no point in explaining to Rodney the teetotaller that we never asked for beer in glass, or what a curse the bottles had been. He just saw them as reinforcement, dozens of times over, of his belief that we were incompetent.

As the evening wore on we ran out of cargo ready for shipping and the previously smooth operation began to falter. We were like an army stretched to its limits, trying desperately to keep the front line provisioned. Our troops were scattered from the Sørensen to the *Shokalskiy*. Joan, Dave Killick, Rod and Mike were back on the ship, handling the incoming cargo. Ted was there also, but to his extreme frustration, he was reduced to a virtual onlooker, having again strained his back. With Rookie and myself crewing on the boats, only five males were left on shore to deal with the remaining gear.

With the cargo stream dwindling to a trickle, the boats were spending longer and longer waiting in Boat Harbour. Our numbers ashore obviously needed reinforcing. Dave Killick and Rod were ferried back in, arriving in a somewhat dispirited state, their ventiles saturated and caked in frozen spray. For Dave, who had thought he was going to die on his very first trip ashore, that second rough ride must have rekindled his worst fears. Rookie and I also stayed behind, while Rodney, Don, Belinda and Phil headed back to the ship for a meal break, taking with them the passengers and Estelle. Showing tourists around the huts was one thing but lumping cargo about was too much for the still 'one-winged' Estelle. Chris Roesler proved to be a real friend. His duties and a quick tour of Cape Denison completed, he set to work as if he'd been with us from the start.

For the next few hours, we laboured furiously to seal the Sørensen and the Apple for winter, clean up the last of our rubbish, pack the remaining gear and tow it all over to Boat Harbour. With every gust that blew slightly stronger than the last, we winced, knowing that the weather could deteriorate at any time, leaving us stranded at the whim of the wind for another day, week or even longer.

Rod was increasingly fretful for our safety. We'd been going non-stop since early morning and tiredness was fast becoming as much of an enemy as the wind. But even in the midst of the most wearisome task or stressful moment, there was cause for humour. Digging for any discarded packaging in the snow bank 'freezer' where we'd stored our meat supplies, Rookie and I chanced on a forgotten cache of 'dead things'—two sides of beef, a slab of pork and several large boxes of prime bacon. Joan later forswore any knowledge of their existence, but for the carnivores among us, the thought of all that denied pleasure was something of a bitter pill to swallow. At least the meat was not going to be wasted, as we'd already made arrangements for most of the surplus food to go to a very appreciative ship's crew.

Unfortunately, where our trusty tents were concerned, there was no funny side. The one big miscalculation in our evacuation was not pulling the tents down until the backloading began. We grossly underestimated how long they would take to drop and instead of the job taking an hour or so for two people, as we had expected, it took probably four times as long and diverted the energies of three of us for most of that time. What's more, it was wearying, hard work, as in latter days, the tents had become so firmly frozen in place that they may easily have withstood the winter.

In the early stages of our stay we were worried about the tents blowing away. As the snow softened and melted around us, the guys loosened and the pegs pulled out. In the battle to keep the tents secure, large stones were heaped onto their skirts and tent pegs replaced with star pickets hammered deep into the snow. But after mid January, as the snows piled up, the névé compacted as a frozen mass. Out of sight, the stones and the ends of the pickets froze in solid ice. When it came time to drop the tents, ice axes, crow bars and sledge hammers were needed. We even thought of retrieving the percussion drill, packed ready for the ship at Boat Harbour.

In retrospect, it's easy to say we should have dropped them earlier, but until the one and only lull, the wind had not blown at less than 40 knots since early the previous Thursday. In such blustery conditions, the already arduous job would have proved a nightmare.

Late in the evening we were just about there. The Sørensen and Apple were shut for the last time and without looking back, we almost ran to Boat Harbour. Rookie and I went over by the most direct route, climbing to the highest point on the ridge near Long Lake. From there, Rookie radioed the ship that we were ready to start loading again.

The signs in the sky were not good though. Just as we dropped into the hut valley, heavy clouds swallowed the last of the evening sun and the penumbra of gold, glinting from Mawson's Hut, was suddenly extinguished. As if we were guests who had overstayed our welcome, the lights were doused, and, while the wind rose once more, we were summarily bid adieu.

The boats soon arrived and the last of the gear from the Weatherhaven was carried to the shore. Dave Gillott had managed to extricate the freezer box full of ice cores with the help of the still toiling chainsaw and it too was loaded. But Rodney was now gravely anxious about our situation and wanted to leave for the ship straight away. The Weatherhaven itself looked destined to stay for a while longer. If it had taken four hours to dig out the other tents, it was going to take just as long to free our work shelter and with the weather rapidly deteriorating, our chances of much more time ashore looked grim. The tide gauge seemed similarly fated. For the time being at least, Rodney intended abandoning the rest of the gear. After a few minutes, Alan arrived on the quad, trailing the last load from the Sørensen, but by that stage human cargo was Rodney's only concern.

The next couple of minutes were excruciating, as Alan and Rodney teetered on the brink of a raging argument. To their mutual credit we were spared the fight and, with Don and Belinda acting as honest brokers, any precipitous actions were averted. Rodney left it to the other two drivers to decide whether to risk further cargo runs. Both Belinda and Don considered the worsening conditions too risky, but they did buy time with Rodney to enable Mawson's Hut to be properly sealed. Dave Gillott scrambled onto the roof of the hut and screwed down the skylight hatches, while Geoff and Paul secured the snow baffles inside and bolted the verandah door. The remaining cargo, plus shovels and crow bars to help dig the Weatherhaven out, were securely cached in the rocks. Lastly, the quad was parked nearby, facing south into the wind.

Drift was now coursing down the hut valley and the waters of Boat Harbour were scuffed with white caps. The searing drift stung our faces, but discomfort was no longer of any concern. We were turning for home, deeply relieved to finally be away. I could almost feel the splash of hot water on my matted hair and the rub of crisp clean sheets on bare legs. With everyone safely aboard, the three heavily laden boats set off together, probably for the last time.

If we were glad to be leaving after two months, what must it have been like for Mawson and the others, finally heading for Australia after two turbulent years. In *The Home of the Blizzard*, Mawson expresses the feeling perfectly:

> We were to live in a land where drift and wind are unknown, where rain fell in refreshing showers, where the sky was blue for long weeks, and where the memories of the past were to fade into a dream—a nightmare?

I lolled on top of boxes and bags in Don's boat, half stupefied with exhaustion and my feet almost trailing in the water. With the wind behind us, I could at last relax and savour the sight of Cape Denison as it slipped into the half light of early winter. First Mawson's Hut, then the Memorial Cross merged with the gloom. Any sign of human endeavour was quickly gone. That mighty landscape that we left behind owed no debt of recognition to our presence or any that had preceded us. We were as inconsequential as a lull in the gale. All my thoughts of that grand place were condensed in a single boat ride of the memory into just two elements: the great arching silhouette of the ice plateau, and the pungent aroma borne on the offshore wind, of countless thousands of Adélie penguins—life as indomitable as the wind itself.

Epilogue: Towards 2011

For all of that Tuesday, the *Akademik Shokalskiy* ploughed on into the gale. Making only three to four knots it was desperately slow going and well before the ship was anywhere near the Mertz Glacier, the closer of the two ice tongues, Captain Vinogradov decided to turn west and run before the wind. Smashing into large swells, the ship's foredeck had begun to ice up badly and water taken over the bow was not draining away. If ice had been allowed to continue accumulating, the ship would have become top heavy and dangerously difficult to manoeuvre. The *Shokalskiy* returned to Commonwealth Bay where the conditions were worse than when we left. The weather forecast faxed to the ship from Casey Station didn't look promising and after waiting a few hours, we reluctantly turned for home at 9 a.m. on Thursday, 5 February, with a 60-knot south-easterly sending us on our way. The quad and the rest of our abandoned gear were there for the next winter at least.

There are so many 'what ifs' and 'maybes' about the final hours ashore at Cape Denison, it's difficult to know exactly where we stumbled. The cooperative spirit between the various parties on the southward journey was sorely missed, but our lack of total preparedness also figures as a cause. Although much of that can be attributed to the appalling conditions during those last days, I feel that because we badly underestimated the severity of late January's weather, our disengagement from Cape Denison was probably in trouble even before the *Shokalskiy* arrived in Commonwealth Bay. And when our chance did come, with our manpower depleted and our numbers scattered between the ship, the landing at Boat Harbour and the Sørensen camp, we simply had too few personnel in any one place for a quick evacuation. With even just one more hour of ship-to-shore operations, I'd probably be claiming close to absolute success. We came so near, yet didn't quite pull off the perfect expedition. For a time after leaving Commonwealth Bay, we were to feel that the edge was missing from our accomplishments.

But to dwell on the single blemish on what was otherwise a quite outstanding string of achievements is to deny the first AAP Mawson's Huts Foundation work party due credit for all that they achieved. Without becoming self-congratulatory, I think our group laboured mightily to successfully complete all it was asked to do and most importantly, everyone returned home safely. Despite our occasional differences, we came through as friends—a formidable accomplishment, given our isolation and the considerable stresses of daily life. To a man and woman, I enjoyed the company of each of my fellows and feel honoured to have been among such a talented group.

And what of my own experiences? Without hesitation, I would say that I left Cape Denison greatly enriched. As an artist with a passion for wild places, the challenges and the rewards could not have been more keenly felt. For some time to come, they'll stay with me as I attempt to make sense of my hard-won impressions under the cool but calm light of studio retrospection. To have lived and worked through the climax and decay of an Antarctic summer was of itself a rare opportunity, but in the circumstances of our project, where we were able to stand in the still vibrant shadow of a remarkable Australian and his comrades, I feel deeply privileged.

There is much that I'll long remember of those turbulent few weeks on the edge of the infinite wilderness. But just as I treasure glimpses of the great themes and mighty forces in Antarctica, there are the small details—the squeak of hard snow underfoot, the trickle of meltwater, the briny, sour smell of washed-up kelp or the distant, haughty cry of the skua—that turn an episode in one's life into a profound experience.

So too with fragments of our daily life. Whether it was an open-air debate with Alan on the meaning of life at 2 a.m. on a breathless evening, or seeing Ted in the swirl of the blizzard slumped with exhaustion over a bogged wheelbarrow full of ice from Mawson's Hut, or watching Dave Killick doing a Captain Hook impression with the bread hook from Joan's kitchen mixer—from the reflective to the serious and the silly, life's most valued times are full of such variety.

Buoyed by the success of the first AAP Mawson's Hut work party, David Jensen and the Foundation are confident of mounting further expeditions to Cape Denison to continue the conservation program and further monitor the condition of the buildings. The summer of 1999–2000 will again see a work party doing battle with the 'home of the blizzard'. In the intervening summer, the *Aurora Australis*, under charter to Italian researchers, will cruise close to the coast of George V Land. The Foundation hopes to have three people on board. With helicopter access to Cape Denison, it's proposed that the small team will be able to retrieve our abandoned gear (including the Weatherhaven and the tide gauge), install some scientific monitoring equipment inside Mawson's Hut and make a general check on the health of the buildings. It may also be possible to shift the Granholm to the Jubilee (Sørensen) Base and remove the old shipping containers and fuel drums.

Even though Mawson's Huts have now been rescued from oblivion, debate will continue among those with professional or statutory interests in the conservation of Australia's Antarctic heritage as to how best to ensure the survival of the huts well past the centenary of the AAE. If anything has changed as a result of our efforts at Cape Denison, it is that the debate can now centre on detail: can a method be devised to properly seal the main hut: should more of the hut be reclad; and if so, when, and how far towards 'restoration' should the interior be taken? And when all debate has died, Mawson's Huts may yet stand resolute to the blizzard in the windiest place on the planet.

Acknowledgments

Many people assisted my passage to Antarctica which in such a short time manifested itself in this book. First and foremost, I'm indebted to David Jensen for his faith and encouragement and to the AAP Mawson's Huts Foundation with David as its chairman for allowing me such a rare and privileged opportunity. To my fellow expeditioners, thank you for your splendid companionship and camaraderie through times both exhilarating and trying. This is your story as much as it is mine.

Sir Edmund Hillary generously found time in his busy life to honour this book with his foreword. To him, my sincere thanks. Among the Expedition's sponsors, I'm extremely appreciative of Fuji Professional (Hanimex Industries) for the supply of film stock and to the Antarctic Division through Kevin Bell for additional rolls and film processing. Thanks also to Les Murphy, Rob Easther, Kalara McGregor, Stephen Martin, Geoff Ashley, Alan Grant, Robyn Grosvenor and Alun Thomas for their help and advice along the way. Permission to quote from *The Home of the Blizzard* and *Mawson's Antarctic Diaries* was kindly given by the University of Adelaide, and the State Library of New South Wales allowed the inclusion of the historic photographs.

I've relished the chance to work once more with Bert Hingley at Hale & Iremonger, and thank him and my editor, Heather Cam, for their encouragement and patience throughout.

Lastly, this book, let alone any of my grand adventures, would remain only a dream if not for Morea Grosvenor.

I would commend the AAP Mawson's Hut Foundation to anyone interested in assisting with the ongoing care and conservation of Australia's Antarctic heritage. The Foundation can be contacted at Locked Bag 21, Grosvenor Place, Sydney 2001.

Alasdair McGregor
Sydney, 1998

Further Reading

Antarctica: Great Stories from the Frozen Continent, Reader's Digest, Sydney, 1985.
Bickel, Lennard, *This Accursed Land*, Macmillan, Melbourne, 1977.
——, *In Search of Frank Hurley*, Macmillan, Melbourne, 1980.
Bowden, Tim, *The Silence Calling, Australians in Antarctica, 1947–97, The ANARE Jubilee History*, Allen & Unwin, Sydney, 1997.
Chester, Jonathan, *Going to Extremes, Project Blizzard and Australia's Antarctic Heritage*, Doubleday, Sydney, 1986.
Fletcher, Harold, *Antarctic Days with Mawson*, Angus & Robertson, Sydney, 1984.
Grenfell Price, A. *The Winning of Australian Antarctica: Mawson's BANZARE Voyages, 1929–31*, Angus & Robertson, Sydney, 1962.
Hurley, Frank, *Argonauts of the South*, Angus & Robertson, Sydney, 1948.
Jacka, Fred and Eleanor (eds), *Mawson's Antarctic Diaries*, Allen & Unwin, Sydney, 1988.
Laseron, Charles, *South with Mawson*, Australasian Publishing Co., 1947.
Martin, Stephen, *A History of Antarctica*, State Library of New South Wales Press, Sydney, 1996.
Mawson, Douglas, *The Home of the Blizzard*, (2 vols), Heinemann, London, 1915. (Abridged edn., Hodder & Stoughton, London, 1930. Facsimile edn., Wakefield Press, Adelaide, 1996.)
McIntyre, Don and Margie (with Peter Meredith), *Two Below Zero: A Year Alone in Antarctica*, Australian Geographic, Sydney, 1996.
Millar, David P., *From Snowdrift to Shellfire, Capt. James Francis (Frank) Hurley (1885–1962)*, David Ell Press, Sydney, 1984.

Members of the AAE Cape Denison party

Lieut. Robert BAGE* (23)^	astronomer, tide gauge keeper and assistant magnetician.
Francis BICKERTON* (22)	air-tractor mechanic.
John CLOSE (40)	assistant collector.
Percy CORRELL (19)	mechanic.
Walter HANNAM (26)	radio operator and mechanic.
Alfred HODGEMAN* (26)	draughtsman, cartographer and artist.
John HUNTER (23)	biologist.
James Francis (Frank) HURLEY (24)	photographer.
Sydney JEFFRYES# (27)	radio operator.
Charles LASERON (25)	collector and taxidermist.
Archie McLEAN* (26)	medical officer and bacteriologist.
Cecil MADIGAN* (23)	meteorologist.
Dr Douglas MAWSON* (30)	expedition leader.
Dr Xavier MERTZ (28)	dog handler and driver, ski instructor.
Herbert MURPHY (32)	storeman.
Lieut. Belgrave NINNIS (23)	dog handler and driver.
Frank STILLWELL (23)	geologist.
Eric WEBB (22)	magnetician.
Leslie WHETTER (29)	surgeon and assistant medical officer.

* Remained for second year. ^ Age at time of joining the expedition. # Second year only.

Glossary

ablation The disappearance of snow or ice through melting or evaporation.

ANARE Australian National Antarctic Research Expeditions. The field operations of the Australian Antarctic Division, Department of Environment, Sport and Territories.

anemometer Device for measuring wind speed.

Apple Prefabricated, dome-shaped demountable field hut used by ANARE. Colour: bright red.

BANZARE British, Australian and New Zealand Antarctic Research Expeditions. Two exploratory cruises led by Sir Douglas Mawson aboard the SY *Discovery* in the summers of 1929–30 and 1930–31.

blue ice Consolidated or glacial ice. Extremely slippery.

brash ice Loose floating fragments of ice broken from ice cliffs, glaciers, etc.

calving The process of an iceberg parting from the continental ice cap or shelves or smaller fragments breaking from icebergs themselves.

crampons Strap on metal spikes worn on the soles of the feet to provide traction when walking or climbing on ice.

crevasse A crack due to pressure and movement in the ice cap or glaciers.

erratic Rocky debris carried from its original position by a glacier.

fast ice Sea ice persisting around the margins of the Antarctic coastline.

glacier tongue Where a glacier disgorges onto the surrounding sea. May float or be grounded.

ice blink The reflected light from sea ice or the ice cap, brightening the sky above. Especially apparent with cloud cover.

ice cap The ice sheet covering Antarctica.

ice foot The persistent apron of ice clinging to the rocky foreshore.

katabatic Wind-generating effect from cold air drainage on the down slope of the polar plateau.

lead A path through the pack ice.

magnetograph Device for measuring fluctuations in the earth's magnetic field.

moraine Rocky debris pushed forward or to the sides of a glacier.

névé Transitional stage between fresh snow and glacial ice.

nunatak An isolated outcrop of rock in the ice cap.

pack ice Floating sea ice. Varies from continuous sheets to widely dispersed pieces.

sastrugi Wind-formed furrows and ridges on a snow or névé surface.

sea ice Frozen sea water.

South Magnetic Pole Point at which the earth's magnetic field is vertical with a magnetised needle inclined to 90°. Corresponding pole in the northern hemisphere. The pole shifts due to the molten nature of the earth's core and has moved about 1000 kilometres to the north-west since early in the century.

tarn A glacier-scoured lake or pond.

ventiles Double-layered, finely woven cotton jacket and overall pants issued to ANARE expeditioners as windproof outer garments.

wind chill Scale measuring the combined chilling effect of temperature and wind speed.